Springer Series in Cognitive and Neural Systems

Volume 13

Series Editor
Vassilis Cutsuridis, School of Computer Science, University of Lincoln,
Lincoln, UK

More information about this series at http://www.springer.com/series/8572

Vassilis Cutsuridis
Editor

Multiscale Models of Brain Disorders

 Springer

Editor
Vassilis Cutsuridis
School of Computer Science
University of Lincoln
Lincoln, UK

ISSN 2363-9105   ISSN 2363-9113   (electronic)
Springer Series in Cognitive and Neural Systems
ISBN 978-3-030-18829-0   ISBN 978-3-030-18830-6   (eBook)
https://doi.org/10.1007/978-3-030-18830-6

This Springer imprint is published by the registered company Springer Nature Switzerland AG.
The registered company address is: Gewerbestrasse 11, 6330 Cham, Switzerland

Preface

Brain disorders have increasing relevance to our societies due to the increasing ageing of the world population and the extremely stressful lives we live in. The aetiology of any brain disorder is usually very complex, and often several hypotheses are needed to explain its pathogenesis. Most of the time, it is experimentally very difficult to understand how the interactions of the various pathways and mechanisms lead to the pathogenesis of a brain disorder and its symptoms. Equally difficult are the various potential routes of cure by drug and electrostimulation therapies. This is mainly because experimental studies are usually carried out to isolate the effects of a single mechanism and do not investigate the interactions of many mechanisms. This leads to a set of results that are conflicting, very difficult to interpret or not integrated in a unified framework.

This limited efficiency of the current experimental methods against the brain disorders has created increasing demands towards the development of other novel therapeutic strategies. Mathematical and computational models have emerged as invaluable tools in resolving such conflicts, because they provide coherent conceptual frameworks for integrating many different spatial and temporal scales and resolutions that allow for observing and experimenting with the neural system as a whole. Computational modellers then have precise control of experimental conditions needed for the replicability of experimental results. Because the process takes place in a computer, the investigator can perform multiple virtual experiments by preparing and manipulating the system in precisely repeatable ways and observe every aspect of the system without interference.

This book provides a series of focused papers on computational models of brain disorders combining multiple levels and types of computation with multiple types of data in an effort to improve understanding, prediction and treatment of brain and mental illness. The book is divided into four thematic areas: (1) movement disorders (e.g. Parkinson's disease), (2) cognitive disorders (e.g. schizophrenia, psychosis, autism, depression), (3) memory disorders (e.g. Alzheimer's disease) and (4) other disorders (e.g. absence epilepsy, anaesthesia). In each thematic area, physiologists and anatomists studying cortical circuits, cognitive neuroscientists studying brain dynamics and behaviour via EEG and functional magnetic resonance imaging (fMRI) and computational neuroscientists using neural modelling techniques are brought together to explore local and large-scale disordered brain dynamics.

This volume is an invaluable resource not only to computational neuroscientists but also to experimental neuroscientists, clinicians, engineers, physicists, mathematicians and all researchers interested in modelling brain disorders. Graduate-level students and trainees in all these fields will find this book an insightful and readily accessible source of information.

Finally, there are many people who I would like to thank for making this book possible. This includes all contributing authors who did a great job. I would like to also thank Paul Roos, my Springer senior editor, and members of the production team, who were consistent source of help and support.

Lincoln, UK   Vassilis Cutsuridis

Contents

Part I   Movement Disorders

**1   A Neuro-computational Model of Pallidal vs. Subthalamic
Deep Brain Stimulation Effect on Synchronization at Tremor
Frequency in Parkinson's Disease**   3
Alekhya Mandali, V. Srinivasa Chakravarthy,
and Ahmed A. Moustafa

**2   Dynamics of Basal Ganglia and Thalamus in Parkinsonian
Tremor**   13
Jan Morén, Jun Igarashi, Osamu Shouno, Junichiro Yoshimoto,
and Kenji Doya

**3   A Neural Mass Model for Abnormal Beta-Rebound
in Schizophrenia**   21
Áine Byrne, Stephen Coombes, and Peter F. Liddle

**4   Basal Ganglio-thalamo-cortico-spino-muscular Model
of Parkinson's Disease Bradykinesia**   29
Vassilis Cutsuridis

**5   Network Models of the Basal Ganglia in Parkinson's Disease:
Advances in Deep Brain Stimulation Through Model-Based
Optimization**   41
Karthik Kumaravelu and Warren M. Grill

**6   Neural Synchronization in Parkinson's Disease on Different
Time Scales**   57
Sungwoo Ahn, Choongseok Park, and Leonid L. Rubchinsky

**7   Obsessive-Compulsive Tendencies and Action Sequence
Complexity: An Information Theory Analysis**   67
Mustafa Zeki, Fuat Balcı, Tutku Öztel, and Ahmed A. Moustafa

Part II   Cognitive Disorders

8   **Cortical Disinhibition, Attractor Dynamics, and Belief
    Updating in Schizophrenia** 81
    Rick A. Adams

9   **Modeling Cognitive Processing of Healthy Controls
    and Obsessive-Compulsive Disorder Subjects
    in the Antisaccade Task** 91
    Vassilis Cutsuridis

10 **Simulating Cognitive Deficits in Parkinson's Disease** 105
    Sébastien Hélie and Zahra Sajedinia

11 **Attentional Deficits in Alzheimer's Disease: Investigating
    the Role of Acetylcholine with Computational Modelling** 113
    Eirini Mavritsaki, Howard Bowman, and Li Su

12 **A Computational Hypothesis on How Serotonin Regulates
    Catecholamines in the Pathogenesis of Depressive Apathy** 127
    Massimo Silvetti, Gianluca Baldassarre, and Daniele Caligiore

13 **Autism Spectrum Disorder and Deep Attractors
    in Neurodynamics** 135
    Włodzisław Duch

Part III   Memory Disorders

14 **Alzheimer's Disease: Rhythms, Local Circuits,
    and Model-Experiment Interactions** 149
    Frances K. Skinner and Alexandra Chatzikalymniou

15 **Using a Neurocomputational Autobiographical Memory Model
    to Study Memory Loss** 157
    Di Wang, Ahmed A. Moustafa, Ah-Hwee Tan, and Chunyan Miao

Part IV   Other Disorders

16 **How Can Computer Modelling Help in Understanding
    the Dynamics of Absence Epilepsy?** 167
    Piotr Suffczynski, Stiliyan Kalitzin, and Fernando H. Lopes da Silva

17 **Data-Driven Modeling of Normal and Pathological Oscillations
    in the Hippocampus** 185
    Ivan Raikov and Ivan Soltesz

18 **Shaping Brain Rhythms: Dynamic and Control-Theoretic
    Perspectives on Periodic Brain Stimulation for Treatment
    of Neurological Disorders** 193
    John D. Griffiths and Jérémie R. Lefebvre

**19 Brain Connectivity Reduction Reflects Disturbed
     Self-Organisation of the Brain: Neural Disorders
     and General Anaesthesia** 207
     Axel Hutt

Index 219

Contributors

Rick A. Adams Institute of Cognitive Neuroscience, University College London, London, UK
Division of Psychiatry, University College London, London, UK

Sungwoo Ahn Department of Mathematics, East Carolina University, Greenville, NC, USA

Fuat Balcı Research Center for Translational Medicine & Department of Psychology, Koç University, Istanbul, Turkey

Gianluca Baldassarre Institute of Cognitive Sciences and Technologies (ISTC-CNR), Rome, Italy

Howard Bowman School of Psychology, University of Birmingham, Birmingham, UK
School of Computing, University of Kent, Canterbury, UK

Áine Byrne Centre for Neural Sciences, New York University, New York, NY, USA

Daniele Caligiore Institute of Cognitive Sciences and Technologies (ISTC-CNR), Rome, Italy

Alexandra Chatzikalymniou Krembil Research Institute, University Health Network, Toronto, ON, Canada
Department of Physiology, University of Toronto, Toronto, ON, Canada

Stephen Coombes Centre for Mathematical Medicine and Biology, School of Mathematical Sciences, University of Nottingham, Nottingham, UK

Vassilis Cutsuridis School of Computer Science, University of Lincoln, Lincoln, UK

Kenji Doya Neural Computation Unit, Okinawa Institute of Science and Technology Graduate University, Okinawa, Japan

Włodzisław Duch Department of Informatics, Faculty of Physics, Astronomy and Informatics, and Neurocognitive Laboratory, Center for Modern Interdisciplinary Technologies, Nicolaus Copernicus University, Toruń, Poland

John D. Griffiths Krembil Centre for Neuroinformatics, Centre for Addiction and Mental Health, Toronto, ON, Canada
Department of Psychiatry, University of Toronto, Toronto, ON, Canada

Warren M. Grill Department of Biomedical Engineering, Duke University, Durham, NC, USA
Department of Electrical and Computer Engineering, Duke University, Durham, NC, USA
Department of Neurobiology, Duke University, Durham, NC, USA
Department of Neurosurgery, Duke University, Durham, NC, USA

Sebastien Hélie Department of Psychological Science, Purdue University, West Lafayette, IN, USA

Axel Hutt Department FE 12 (Data Assimilation), Deutscher Wetterdienst – German Meteorological Service, Research and Development, Offenbach am Main, Germany

Jun Igarashi Computational Engineering Applications Unit, RIKEN, Wako, Japan

Stiliyan Kalitzin Foundation Epilepsy Institute in The Netherlands (SEIN), Heemstede, The Netherlands
Image Sciences Institute, University Medical Center Utrecht, Utrecht, The Netherlands

Karthik Kumaravelu Department of Biomedical Engineering, Duke University, Durham, NC, USA

Jérémie R. Lefebvre Krembil Research Institute, University Health Network, Toronto, ON, Canada
Department of Mathematics, University of Toronto, Toronto, ON, Canada

Peter F. Liddle Institute of Mental Health, School of Medicine, University of Nottingham, Nottingham, UK

Fernando H. Lopes da Silva Swammerdam Institute for Life Sciences, Center of Neuroscience, University of Amsterdam, Amsterdam, The Netherlands
Department of Bioengineering, Instituto Superior Técnico, Lisbon Technical University, Lisbon, Portugal

Alekhya Mandali Department of Psychiatry, School of Clinical Medicine, University of Cambridge, Cambridge, UK

Eirini Mavritsaki Department of Psychology, Birmingham City University, Birmingham, UK
School of Psychology, University of Birmingham, Birmingham, UK

Chunyan Miao Joint NTU-UBC Research Centre of Excellence in Active Living for the Elderly, Nanyang Technological University, Singapore
School of Computer Science and Engineering, Nanyang Technological University, Singapore

Jan Morén Scientific Computing and Data Analysis Section, Okinawa Institute of Science and Technology Graduate University, Okinawa, Japan

Ahmed A. Moustafa School of Social Sciences and Psychology & Marcs Institute for Brain and Behaviour, Western Sydney University, Sydney, New South Wales, Australia

Tutku Öztel Research Center for Translational Medicine, Koç University, Istanbul, Turkey

Choongseok Park Department of Mathematics, North Carolina A&T State University, Greensboro, NC, USA

Ivan Raikov Department of Neurosurgery, Stanford University, Stanford, CA, USA

Leonid L. Rubchinsky Department of Mathematical Sciences, Indiana University-Purdue University Indianapolis, Indianapolis, IN, USA
Stark Neurosciences Research Institute, Indiana University School of Medicine, Indianapolis, IN, USA

Zahra Sajedinia Department of Psychological Science, Purdue University, West Lafayette, IN, USA

Osamu Shouno Honda Research Institute Japan Co., Ltd, Wako, Japan

Massimo Silvetti Department of Experimental Psychology, Ghent University, Ghent, Belgium

Frances K. Skinner Krembil Research Institute, University Health Network, Toronto, ON, Canada
Department of Medicine (Neurology) and Physiology, University of Toronto, Toronto, ON, Canada

Ivan Soltesz Department of Neurosurgery, Stanford University, Stanford, CA, USA

V. Srinivasa Chakravarthy Department of Biotechnology, Indian Institute of Technology, Madras, Chennai, India

Piotr Suffczynski Department of Biomedical Physics, Institute of Experimental Physics, University of Warsaw, Warsaw, Poland

Li Su Department of Psychiatry, University of Cambridge, Cambridge, UK
Sino-Britain Centre for Cognition and Ageing Research, Southwest University, Chongqing, China

Ah-Hwee Tan Joint NTU-UBC Research Centre of Excellence in Active Living for the Elderly, Nanyang Technological University, Singapore
School of Computer Science and Engineering, Nanyang Technological University, Singapore

Di Wang Joint NTU-UBC Research Centre of Excellence in Active Living for the Elderly, Nanyang Technological University, Singapore

Junichiro Yoshimoto Mathematical Informatics Laboratory, Nara Institute of Science and Technology, Ikoma, Japan

Mustafa Zeki Department of Mathematics, College of Engineering and Technology, American University of the Middle East, Kuwait

Part I
Movement Disorders

Chapter 1
A Neuro-computational Model of Pallidal vs. Subthalamic Deep Brain Stimulation Effect on Synchronization at Tremor Frequency in Parkinson's Disease

Alekhya Mandali, V. Srinivasa Chakravarthy, and Ahmed A. Moustafa

Abstract Parkinson's disease is a neurodegenerative disorder, associated with different motor symptoms including tremor, akinesia, bradykinesia, rigidity as well as gait and speech impairments. Previously, we have presented a neurobiologically detailed neuro-computational model simulating the basal ganglia functioning as well as the effects of subthalamic deep brain stimulation on action section (Mandali A, Chakravarthy VS, Rajan R, Sarma S, Kishore A, Front Physiol 7:585, 2016; Mandali A, Rengaswamy M, Chakravarthy S, Moustafa AA, Front Neurosci 9:191, 2015). In the current study, we extend our prior model by including thalamic and cortical neurons and compare the effect of subthalamic and pallidal stimulation on tremor in terms of oscillations within STN and GPi and subsequently their effect on the cortex. In agreement with existing experimental studies, our model shows that subthalamic stimulation is more effective at reducing the tremor power than the pallidal stimulation. Our model provides a mechanistic explanation for such comparative results.

Keywords Parkinson's disease · Deep brain stimulation · Izhikevich spiking neuron · Tremor · Sub thalamic nucleus · Globus pallidus

A. Mandali
Department of Psychiatry, School of Clinical Medicine, University of Cambridge, Cambridge, UK

V. Srinivasa Chakravarthy (✉)
Department of Biotechnology, Indian Institute of Technology, Madras, Chennai, India
e-mail: schakra@iitm.ac.in

A. A. Moustafa
School of Social Sciences and Psychology & Marcs Institute for Brain and Behaviour, Western Sydney University, Sydney, New South Wales, Australia
e-mail: a.moustafa@westernsydney.edu.au

© Springer Nature Switzerland AG 2019
V. Cutsuridis (ed.), *Multiscale Models of Brain Disorders*, Springer Series in Cognitive and Neural Systems 13, https://doi.org/10.1007/978-3-030-18830-6_1

1.1   Introduction

Parkinson's disease (PD) is a neurodegenerative disorder associated with a set of motor symptoms such as tremor, bradykinesia, rigidity and gait abnormalities [27]. Apart from motor abnormalities, PD patients also suffer from non-motor problems such as autonomic dysfunction, impaired cognitive functioning and speech abnormalities [7, 8, 19]. Initially, pharmacological treatments with combination of drugs such as levodopa (a precursor of DA), dopamine agonists (DAA) such as ropinirole and pramipexole and MAO inhibitors [21] are prescribed. However, 50–80% of PD patients develop dyskinesias within 2–5 years of L-DOPA treatment [1, 9, 11] or show cognitive deficits such as impulsivity due to the administration of DAA [22, 25]. In conditions where the 'ON time' (patients on medication) is clinically not effective or the patients show refractory PD symptoms, clinicians suggest a surgical technique called deep brain stimulation (DBS). It involves implanting an electrode in deep structures of the brain such as ventralis intermedius (Vim) of the thalamus, globus pallidus internus (GPi) and subthalamic nucleus (STN) [4, 12, 24] among which GPi and STN are a part of the basal ganglia. Among these targets, STN and GPi stimulation are widely chosen as neural targets [3] due to their higher therapeutic benefit.

We have conducted a literature review on the comparison between the STN and GPi with respect to symptomatic relief, which has shown differential effects on motor, cognitive and other domains (Table 1.1). For example, it has been observed that the symptomatic relief from bradykinesia and rigidity through STN

Table 1.1 Studies comparing the effect of STN vs. GPi stimulation on various PD symptoms

S. no.	Symptom	The effect of stimulation
1	Resting tremor	STN is superior to GPi [2]
2	Bradykinesia, rigidity (responsive to dopaminergic medication)	Bilateral STN stimulation improves bradykinesia more than bilateral and unilateral GPi stimulation [2, 26]
3	Gait and postural instability	Bilateral GPi stimulation subjects showed improvement in stand-walk-sit test and in gait [26]
4	Dystonia	Both STN and GPi have equal advantage [6]
5	Reduction in medication	Reduction in L-DOPA dosage by 38% for STN stimulation [6]
6	On-off fluctuations	GPi effect >STN effect [6]
7	Dyskinesias	GPi stimulation has dyskinesia suppression up to 89% than 62% from STN stimulation [26]
8	Off-time motor symptom	No suggestive advantage of any target stimulation [26]
9	Cognition	A higher decline in cognitive levels of subjects with STN stimulation compared to GPi [26]
10	Long-term medication and management	GPi stimulation was observed to have an advantage in the flexibility of long-term medication and management of the DBS lead [26]

stimulation is higher than GPi, whereas relief from dyskinesias is better via GPi stimulation compared to STN stimulation. On the cognitive front, the side effects in terms of impulsivity due to STN stimulation were observed to be higher than GPi. Irrespective of its wide clinical usage, the mechanism by which DBS is effective is still unclear. It is in these scenarios that computational modelling comes handy, where one can study the effect of stimulation at the level of neural activity, symptoms and behaviour.

Earlier, we built a spiking network model of the basal ganglia to study the neural activity and cognitive functioning (using decision-making tasks) of PD patients with and without medication and stimulation. We focused on how STN stimulation could modulate the neural activity which may induce impulsivity. The results from the model suggested that the electrode position within STN and the amplitude of the current could significantly vary the behaviour of the patient which could induce impulsive behaviour [14, 15, 17, 18]. We also observed from simulation that an antidromic activation of the GPe neuron could also affect the behaviour by modulating the STN activity.

In the current chapter, we have extended the same computational model [16] by including Izhikevich neuron-based thalamic and cortical neurons and studied the neural activity in the basal ganglia (BG) and cortical neurons and their relation to tremor with and without external stimulation.

1.2 Effect of Pallidal vs. Subthalamic Stimulation on Tremor

Although the effect of medication and stimulation on the motor symptoms of PD have been extensively studied experimentally [2, 3], most of the computational studies have either concentrated on understanding the dynamics of the cortical-subcortical structures and their interaction in terms of synchrony, firing pattern and oscillations [5] or simulated the arm dynamics by incorporating abstract but not detailed simulations of neural structures [20].

We studied the effect of deep brain stimulation on the frequency content in STN and GPi neurons around tremor frequency using our previously published model ([16]). The network model of cortico-BG neurons was built using two-variable Izhikevich spiking neurons where each nucleus was modelled as a lattice with $(= 50 \times 50)$ of neurons. All the neurons are connected in one-one fashion with the striatum, globus pallidus externa (GPe) and GPi being inhibitory and thalamus, STN and cortex being excitatory. Each GPi neuron receives both glutamatergic projections from STN and GABAergic projections from D1R-expressing striatal MSNs. The information flow from cortex enters both D1 and D2 striatum and later follows the direct

(continued)

and indirect pathways finally reaching the thalamus which further projects back to the cortex (Fig. 1.1). Equations related to the Izhikevich spiking neuron model are described as

$$\frac{dv_{ij}^x}{dt} = 0.04\left(v_{ij}^x\right)^2 + 5v_{ij}^x - u_{ij}^x + 140 + I_{ij}^x + I_{ij}^{syn} \tag{1.1}$$

$$\frac{du_{ij}^x}{dt} = a\left(bv_{ij}^x - u_{ij}^x\right) \tag{1.2}$$

$$\text{if } v_{ij}^x \geq v_{peak} \left\{ \begin{array}{l} v_{ij}^x \leftarrow c \\ u_{ij}^x \leftarrow u_{ij}^x + d \end{array} \right\} \tag{1.3}$$

where v_{ij}^x = membrane potential, u_{ij}^x = membrane recovery variable, I_{ij}^{Syn} = total synaptic current received, I_{ij}^x = external current applied to neuron x at location (i, j) and v_{peak} = maximum voltage set to neuron (+30mv) with x being STN or GPe or GPi neuron. The values for the parameters a, b, c and d for STN, GPi and GPe are given in ([16]). The values for striatum, thalamus and cortex are given below (a_{str} = 0.02, b_{str} = 0.2, c_{str} = −65, d_{str} = 8, a_{th} = 0.002, b_{th} = 0.25, c_{th} = −65, d_{th} = 0.05, a_{ctx} = 0.02, b_{ctx} = 0.2, c_{th} = −65, d_{ctx} = 8).

Fig. 1.1 Pictorial representation of the model with key BG nuclei such as the striatum, STN, GPe and GPi. All synaptic connections are GABAergic for all nuclei except for the STN, thalamus and cortex which are glutamatergic

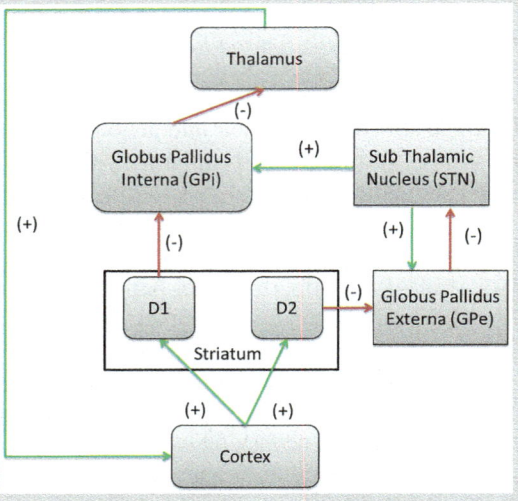

The synaptic connectivity between all the nuclei (cortex, thalamus, GPi, striatum, STN and GPe) was modelled as

$$\tau_{Recep} * \frac{dh_{ij}^{x \to y}}{dt} = -h_{ij}^{x \to y}(t) + S_{ij}^{x}(t) \qquad (1.4)$$

$$I_{ij}^{x \to y}(t) = W_{x \to y} * h_{ij}^{x \to y}(t) * \left(E_{Recep} - V_{ij}^{y}(t) \right) \qquad (1.5)$$

where τ_{Recep} is the decay constant for synaptic receptor, E_{Recep} is the receptor-associated synaptic reversal potential (Recep = AMPA/GABA/NMDA), S_{ij}^{x} is the spiking activity of neuron 'x' at time 't', $h_{ij}^{x \to y}$ is the gating variable for the synaptic current from 'x' to 'y', $W_{x \to y}$ is the synaptic weight from neuron 'x' to 'y' and V_{ij}^{y} is the membrane potential of the neuron 'y' for the neuron (cortex/thalamus/GPi/striatum/STN/GPe) at the location (i,j).

The effect of dopamine on the glutamatergic current from the cortex on to D1 and D2 striatal neurons was modelled as

$$I_{ij}^{Ctx \to D1} = \left(I_{ij}^{NMDA \to D1} + I_{ij}^{AMPA \to D1} \right) * cD1 \qquad (1.6)$$

$$I_{ij}^{Ctx \to D2} = \left(I_{ij}^{NMDA \to D2} + I_{ij}^{AMPA \to D2} \right) * cD2 \qquad (1.7)$$

where $cD1 = \frac{A_{D1}}{1+\exp(-\lambda^{Str}*(DA-1))}$, $cD2 = \frac{A_{D2}}{1+\exp(\lambda^{Str}*DA)}$, DA= dopamine level (0.1–0.9), $\lambda=$, $A_{D1}=$ 30 and $A_{D2}=10$.

Based on the experimental results that dopaminergic receptors (D2) are present on STN and GPe neurons which modulated their synaptic strengths, we modelled the effect of DA on STN-GPe synaptic strength as

$$W_{x \to y} = (1 - cd2 * DA) * w_{x \to y} \qquad (1.8)$$

where the synapses are GPe→STN and STN→GPe. A similar method for DA-dependent synaptic modulation on striatal neurons was used in [10].

We first simulated the PD condition by keeping DA low (= 0.1) (parameter 'DA' in Eqs. 1.6, 1.7 and 1.8) in the model and calculated the oscillatory frequency using fast Fourier transform (FFT) which showed a peak at around 8 Hz (Fig. 1.2c)

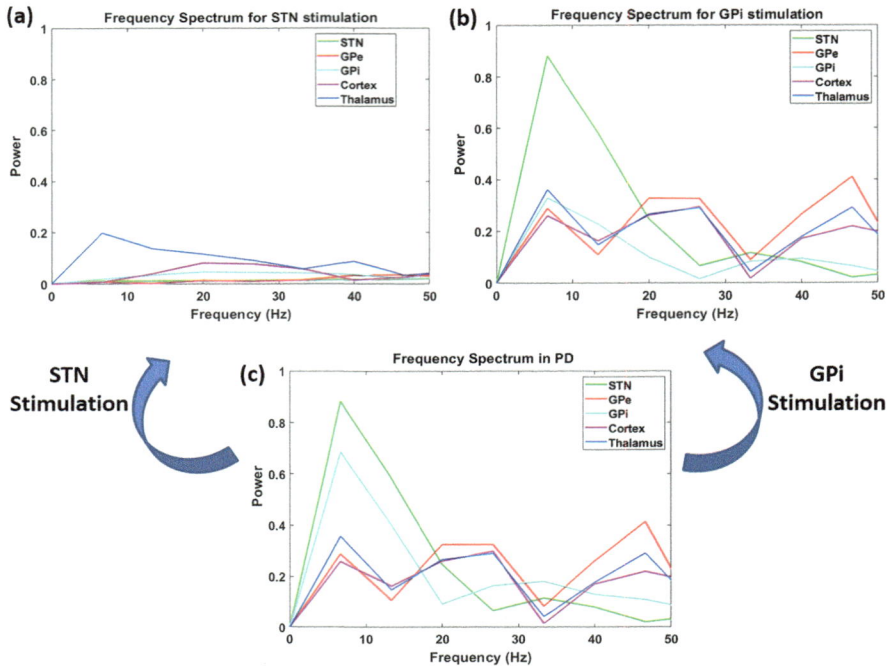

Fig. 1.2 Power in STN and GPi neurons with and without external stimulation. (**a**) Frequency content in STN and GPi in the PD condition and when STN was stimulated; (**b**) frequency content in STN and GPi in the PD condition when GPi was stimulated; and (**c**) frequency content in STN and GPi in PD condition without any external stimulation

in the BG, thalamic and cortical neurons. To understand the effect of stimulation, we individually stimulated STN and GPi neurons with monophasic current (frequency = 130 Hz, amplitude = 120 pA, pulse width = 100 µS). The cortico-thalamic weight was kept at value of (=20) which gave rise to the oscillations in the thalamus and cortex.

The stimulation of STN neurons resulted in a sharp dip in the power of oscillations (Fig. 1.2a) in both BG and cortical neurons. This pattern was absent during GPi stimulation, i.e. only the power of oscillations in GPi was decreased but was unaltered in the STN as well as the cortex (Fig. 1.2b). This result only further reinforces the earlier theory of STN-GPe circuit to being the potential tremor frequency generator [23]. As the STN activity peaks up at around tremor frequency (4–8 Hz), the GPi activity follows the glutamatergic input pattern from STN, and this behaviour is further reverberated into the

(continued)

loop through the thalamus into the cortex. Once the external DBS stimulation is applied to the STN neurons, the synchronous oscillatory activity within itself is disrupted, which further halts the propagation of the oscillations in the entire loop. The cortex, now free from the forced oscillations of STN, resumes its spontaneous activity which has a mean frequency of around 20 Hz as observed in Fig. 1.2a. This spontaneous firing rate which was earlier masked by the tremor frequency oscillations is prominent only after the STN but not GPi stimulation. This is believed to be due to the non-suppression of oscillations in STN neurons during stimulation of GPi neurons.

The above result shows that STN stimulation is effective in suppressing the oscillations around the tremor frequency within itself and GPi through its glutamatergic projections. The presence of intact oscillatory behaviour in STN neurons and only a slight decrease in oscillatory power during GPi stimulation might be a plausible reason for STN being a preferred target for tremor treatment (as indicated in Table 1.1).

We have also tested the role of thalamo-cortical weight in the 'resonating' effect within the cortico- BG loop. At low DA conditions (DA = 0.1), a higher thalamo-cortical weight is believed to induce the resonating property within the loop, i.e. the cortical input to striatum is backpropagated to the cortex through the thalamus which is otherwise absent in healthy (low thalamo-cortical connection) conditions. We simulated three (low ($w = 2$), medium ($w = 10$) and high ($w = 20$)) thalamo-cortical weight conditions and observed for the presence of such a replay within the cortical neurons. The weight influences the glutamatergic current from the thalamus to the cortex which was modelled as a combination of both AMPA and NMDA currents, similar to Eqs. 1.6 and 1.7.

The cortex in the spiking model was stimulated by a Gaussian pulse at a specific time point, one of the standard ways to replicate the synaptic current propagation among neuronal networks [13]. The presence of resonance or backpropagation through the thalamus to the cortex is observed in terms of constant/increase spiking activity in cortex. Figure 1.3 shows the effect of thalamo-cortical weight on the reverberation in cortical neurons. In low-weight condition ($w = 2$), the spiking cortical neuron activity slowly decays down over time (Fig. 1.3c). But on further increase of the connection strength ($w = 10$), the rate of decay in the spiking activity slows down which is observed as the constant spiking activity (Fig. 1.3b), and at higher connection strengths ($w = 20$, (Fig. 1.3a)), an increase in cortical activity over time is observed which is believed to be due to the back propagation property of the loop. These simulations indicate that thalamo-cortical connection strength

(continued)

might play a crucial role in the propagation/reverberation of the activity within the cortico-BG loop.

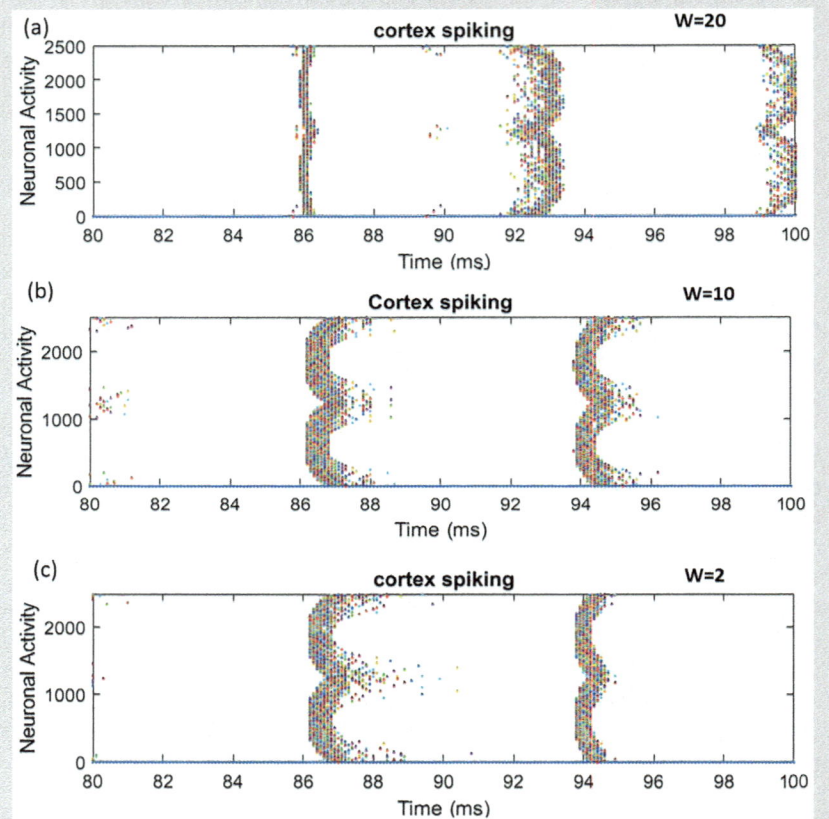

Fig. 1.3 Change in cortical activity as the strength of thalamo-cortical weight is decreased after a Gaussian pulse at time = 85 milliseconds (ms): (**a**) for high weight (w = 20), the cortical activity is maintained, (**b**) for medium weight (w = 10), the cortical activity starts to decrease after a period of time and (**c**) at lower weight (w = 2), a further decrease in cortical activity is observed

One of the studies that attempted to understand the effect of stimulation on motor symptoms was conducted by [20]. The authors simulated the cortico-BG dynamics using the Contreras-Vidal and Stelmach closed-loop model and combined it with the Grossberg's VITE model which calculates the difference between the desired target and current position to determine the direction of motion. The simulation results showed that stimulation to STN helped to restore the normal activity through either stimulation-induced inhibition of STN or partial synaptic failure of efferent projections, or excitation of inhibitory afferent axons.

Though the model by Moroney et al. [20] was able to shed some light on the possible mechanism of DBS action, each of the neurons in the basal ganglia and cortex were modelled as rate coded and lacked the detailed spiking dynamics. Furthermore, there was no explicit arm in the model which could replicate tremor or bradykinesia. Based on our current results (Fig. 1.2), we intend to extend the current spiking neuron model by integrating it to a biologically realistic arm and simulate the PD motor symptoms by altering the dynamics of the neural structures. The effect of STN/GPi stimulation at symptomatic as well as individual neuron level will be further studied.

References

1. Ahlskog JE, Muenter MD (2001) Frequency of levodopa-related dyskinesias and motor fluctuations as estimated from the cumulative literature. Mov Disord 16(3):448–458
2. Anderson VC, Burchiel KJ, Hogarth P, Favre J, Hammerstad JP (2005) Pallidal vs subthalamic nucleus deep brain stimulation in Parkinson disease. Arch Neurol 62(4):554–560
3. Andrade P, Carrillo-Ruiz JD, Jiménez F (2009) A systematic review of the efficacy of globus pallidus stimulation in the treatment of Parkinson's disease. J Clin Neurosci 16(7):877–881
4. Benabid A, Benazzouz A, Hoffmann D, Limousin P, Krack P, Pollak P (1998) Long-term electrical inhibition of deep brain targets in movement disorders. Mov Disord 13(S3):119–125
5. Beuter A, Modolo J (2009) Delayed and lasting effects of deep brain stimulation on locomotion in Parkinson's disease. Chaos: Interdisc J Nonlinear Sci 19(2):026114
6. Castrioto A, Moro E (2013) New targets for deep brain stimulation treatment of Parkinson's disease. Expert Rev Neurother 13(12):1319–1328
7. Chaudhuri KR, Healy DG, Schapira AH (2006) Non-motor symptoms of Parkinson's disease: diagnosis and management. Lancet Neurol 5(3):235–245
8. Chaudhuri KR, Odin P, Antonini A, Martinez-Martin P (2011) Parkinson's disease: the non-motor issues. Parkinsonism Relat Disord 17(10):717–723
9. Fahn S (2000) The spectrum of levodopa-induced dyskinesias. Ann Neurol 47(4 Suppl 1):S2–S9. discussion S9–11
10. Humphries MD, Lepora N, Wood R, Gurney K (2009) Capturing dopaminergic modulation and bimodal membrane behaviour of striatal medium spiny neurons in accurate, reduced models. Front Comput Neurosci 3:26
11. Kishore A, Meunier S, Popa T (2014) Cerebellar influence on motor cortex plasticity: behavioral implications for Parkinson's disease. Front Neurol 5:68
12. Lozano AM, Dostrovsky J, Chen R, Ashby P (2002) Deep brain stimulation for Parkinson's disease: disrupting the disruption. Lancet Neurol 1(4):225–231
13. Lukasiewicz PD, Werblin FS (1990) The spatial distribution of excitatory and inhibitory inputs to ganglion cell dendrites in the tiger salamander retina. J Neurosci 10:210–221
14. Mandali A, Chakravarthy VS (2015a) A computational basal ganglia model to assess the role of STN-DBS on Impulsivity in Parkinson's disease. Paper presented at the Neural Networks (IJCNN), 2015 International Joint Conference on Neural Networks (IJCNN)
15. Mandali A, Chakravarthy VS (2015b) A spiking network model of basal ganglia to study the effect of dopamine medication and STN-DBS during probabilistic learning task. BMC Neurosci 16(Suppl 1):P25

16. Mandali A, Rengaswamy M, Chakravarthy S, Moustafa AA (2015) A spiking Basal Ganglia model of synchrony, exploration and decision making. Front Neurosci 9:191
17. Mandali AC, Chakravarthy VS (2016) Probing the role of DBS electrode position and antidromic activation on impulsivity using a computational model of Basal Ganglia. Front Hum Neurosci 10:450. https://doi.org/10.3389/fnhum.2016.00450
18. Mandali A, Chakravarthy VS, Rajan R, Sarma S, Kishore A (2016) Electrode position and current amplitude modulate impulsivity after subthalamic stimulation in Parkinsons disease— a computational study. Front Physiol 7:266
19. Merello M (2007) Non-motor disorders in Parkinson's disease. Rev Neurol 47(5):261–270
20. Moroney R, Heida C, Geelen J (2008) Increased bradykinesia in Parkinson's disease with increased movement complexity: elbow flexion–extension movements. J Comput Neurosci 25(3):501
21. Nomoto M, Iwata S-i, Kaseda S (2001) Pharmacological treatments of Parkinson's disease. Nihon yakurigaku zasshi Folia Pharmacol Jpn 117(2):111–122
22. Piray P, Zeighami Y, Bahrami F, Eissa AM, Hewedi DH, Moustafa AA (2014) Impulse control disorders in Parkinson's disease are associated with dysfunction in stimulus valuation but not action valuation. J Neurosci 34(23):7814–7824
23. Plenz D, Kital ST (1999) A basal ganglia pacemaker formed by the subthalamic nucleus and external globus pallidus. Nature 400(6745):677–682
24. Volkmann J (2004) Deep brain stimulation for the treatment of Parkinson's disease. J Clin Neurophysiol 21(1):6–17
25. Voon V, Reynolds B, Brezing C, Gallea C, Skaljic M, Ekanayake V et al (2010) Impulsive choice and response in dopamine agonist-related impulse control behaviors. Psychopharmacology 207(4):645–659
26. Williams NR, Foote KD, Okun MS (2014) Subthalamic nucleus versus globus pallidus internus deep brain stimulation: translating the rematch into clinical practice. Mov Disord Clin Pract 1(1):24–35
27. Xia R, Mao Z-H (2012) Progression of motor symptoms in Parkinson's disease. Neurosci Bull 28(1):39–48

Chapter 2
Dynamics of Basal Ganglia and Thalamus in Parkinsonian Tremor

Jan Morén, Jun Igarashi, Osamu Shouno, Junichiro Yoshimoto, and Kenji Doya

Abstract Although beta-range oscillation is observed in the basal ganglia (BG) of Parkinson's disease (PD) patients, its causal role in Parkinson's tremor has been controversial because the PD tremor is in much lower frequency range. In order to explore the dynamic interaction between the BG and their downstream, we built a spiking neuron model of the BG-thalamocortical (TC) network at the scale of a rat brain. The model of the BG-TC circuit reproduced normal, asynchronous firing state and Parkinson-like state with 15 Hz beta-range oscillation in the BG and 4–8 Hz oscillation in the thalamus and the cortex, which is consistent with the PD tremor frequency. Simulation results show that hyperpolarizing current in the thalamic neurons and synchronous burst input from the BG are essential for subharmonic resonant response of the thalamic network in generating Parkinsonian tremor.

This work was supported by HPCI Strategic Program for Computational Life Science and Application in Drug Discovery and Medical Development by MEXT.

J. Morén
Scientific Computing and Data Analysis Section, Okinawa Institute of Science and Technology Graduate University, Okinawa, Japan
e-mail: jan.moren@oist.jp

J. Igarashi
Computational Engineering Applications Unit, RIKEN, Wako, Japan
e-mail: jigarashi@riken.jp

O. Shouno
Honda Research Institute Japan Co., Ltd, Wako, Japan
e-mail: shouno@jp.honda-ri.com

J. Yoshimoto
Mathematical Informatics Laboratory, Nara Institute of Science and Technology, Ikoma, Japan
e-mail: juniti-y@is.naist.jp

K. Doya (✉)
Neural Computation Unit, Okinawa Institute of Science and Technology Graduate University, Okinawa, Japan
e-mail: doya@oist.jp

Keywords Parkinson's disease · Tremor · Beta-range oscillation · Sub-harmonic resonance · Basal ganglia · Thalamus · Spiking neural network · Computational modeling

2.1 Introduction

A major symptom of Parkinson's disease (PD) is tremors in the limbs. The basal ganglia (BG) are the main locus of the disorder, and 15–30 Hz beta-range oscillation is observed in the BG of PD patients and model animals [7]. However, typical PD limb tremors are in 4–8 Hz range. If the beta-range oscillation in BG is responsible for the tremor, there should be some nonlinear dynamics to cause sub-harmonic resonance at the levels of the thalamus, the cortex, the spinal cord, or the musculoskeletal system. The thalamus is a major output target of the BG, and the thalamic neurons are known to oscillate synchronously at 4–6 Hz frequency range during sleep [1]. Here we consider a hypothesis that PD tremor is caused by the dynamic interaction of the BG and the thalamus [3, 10].

In order to test the hypothesis, we constructed spiking neural network models of the BG, the thalamus, and the motor cortex (Fig. 2.1). The models take into account topographic organization in those areas and are scaled to the size of the rat brain.

2.2 Spiking Neural Network Models

We follow the BG model by Shouno et al. [11, 12], in which 8–20 Hz beta-range oscillation originates from the excitatory-inhibitory feedback interaction of the subthalamic nucleus and the external globus pallidus [13].

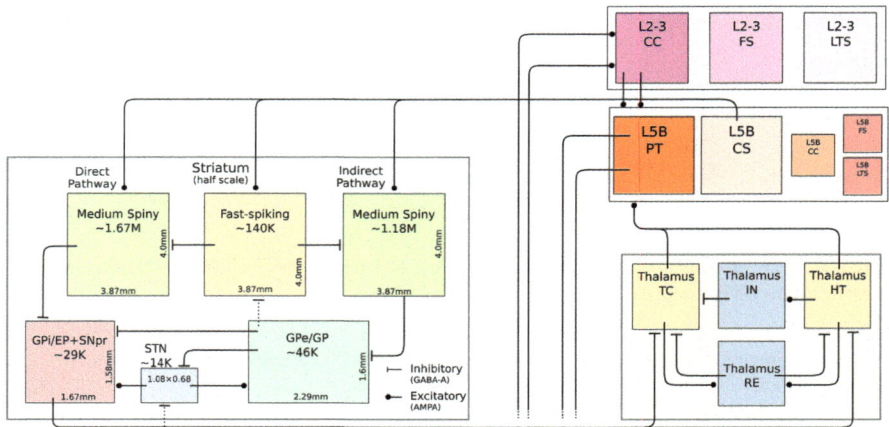

Fig. 2.1 The organization of the basal ganglia-thalamocortical spiking neural network model

We use the neuron models from [12] and lay them out in a spatial manner, with stochastic, overlapping connection patterns derived from experimental data, rather than disjoint "channels" as assumed in [11].

We estimated the spatial dimensions of the BG from the literature [9] and constructed rectangular 2-D surfaces with the width, height, and area same as the rat and the depth to match the volume. We populated each area with the reported number of neurons in an even grid (Table 2.1).

The BG model consists of six neuron types (Fig. 2.1, left): the striatum with fast spiking neurons (FS) and medium spiny neurons projecting to the direct pathway (MS D1) and the indirect pathway (MS D2); the subthalamic nucleus (STN); the external globus pallidus (GPe, called GP in the rat); and the internal globus pallidus (GPi, called endopeduncular nucleus (EP) in the rat) and the substantia nigra pars reticulata (SNpr) combined. The striatum receives inputs from the cortical layer 5B, and GPi/EP+SNpr sends output to the thalamus. The specifications of connections within and between these neuron types are shown in Table 2.2.

The thalamic network consists of four neuron types (Fig. 2.1, right bottom): 800 thalamocortical (TC) neurons, 1200 high-threshold (HT) neurons, 800 interneurons (IN), and 800 reticular (RE) neurons. Neurons are conductance-based models based

Table 2.1 The structural parameters of the BG network model

Area	Size (mm)	Neuron type	Neuron number
Striatum	3.78*4.0	FS	0.14 M
		MSN D1	1.7 M
		MSN D2	1.1 M
STN	1.08*1.68	STN	14 K
GPe	2.29*1.6	GPe	46 K
GPi + SNpr	1.67*1.58	GPi	29 K

Table 2.2 The connection parameters of the BG network model. The weights in ( ) are those for PD state. Connection radius shows the target radius, except those with * showing the source radius

From	To	Weight	Delay (ms)	Probability	Radius (μm)
Cortex 5B	MS D1	5.0	5.0	1.0	32.5
Cortex 5B	MS D2	10.0	5.0	1.0	32.5
Cortex 5B	FS	5.0	5.0	1.0	32.5
FS	MS D1	0.12	1.0	0.5	32.5
FS	MS D2	0.08	1.0	0.5	32.5
MS D1	GPi	−0.1 (−0.07)	1.0	1.0	47.25
MS D2	GPe	−0.12 (−0.5)	1.0	1.0	48
GPe	GPe	−0.01 (−0.03)	1.0	0.05	*96–240
GPe	STN	−4.5 (−6.38)	5.0	0.13	*48
GPe	GPi	−0.1	5.0	0.13	*48
STN	GPe	0.7 (2.0)	5.0	0.22	*21.75
STN	GPi	0.7	5.0	0.22	*21.75

Table 2.3 The connection parameters of the thalamic network model. The weights in ( ) are those for PD state. Connection radius shows the target radius

From	To	Weight	Delay (ms)	Probability	Radius (μm)
GPi	TC	−0.15 (−0.1)	5.0	1.0	33.6
GPi	HT	−0.03 (−0.02)	5.0	1.0	33.6
TC	RE	0.25	1.0	1.0	200
HT	IN	0.5	1.0	1.0	200
HT	RE	0.25	1.0	1.0	200
IN	TC	−0.025	1.0	1.0	200
RE	RE	−0.0125	1.0	1.0	200
RE	TC	−0.01	1.0	1.0	200
RE	HT	−0.015	1.0	1.0	200

on [1]. When hyperpolarized, TC neurons show low-threshold spike (LTS) bursts. The connections to and within the thalamic network model are shown in Table 2.3.

The cortical network model consists of eight neuron types (Fig. 2.1, right top): 44k cortico-cortical (CC), 5.5k fast spiking (FS), and 5.5k low-threshold spiking (LTS) neurons in layers 2–3, and 18.2k pyramidal tract (PT), 9.1k corticostriatal (CS), 9.1k CC, 4.55k FS, and 4.55k LTS neurons in layer 5B. All cortical neurons are conductance-based integrate-and-fire model with alpha-shaped synaptic inputs [6]. The entire TC network contains about 180,000 neurons.

PD state is simulated by (1) weakened MS D1 connections to GPi and strengthened MS D2 connections to GPe; (2) strengthened GPe connections to STN and GPi and within GPe; and (3) weakened GPi connections to TC and HT, as shown in ( ) in Tables 2.2 and 2.3.

2.3 Simulation Results

The BG and TC models were implemented using the NEST simulator [5] and connected by the MUSIC framework [2] to run together on highly parallel computers, including RIKEN's K supercomputer. The combined model reproduced a normal state with asynchronous firing and a PD-like state with beta-range burst oscillation in STN, GPe, and GPi [8].

2.3.1 Oscillatory Properties of the Thalamic Network

Here we examine the intrinsic oscillatory property of the thalamus as a possible source of Parkinsonian tremor. The TC neurons have hyperpolarizing bias current,

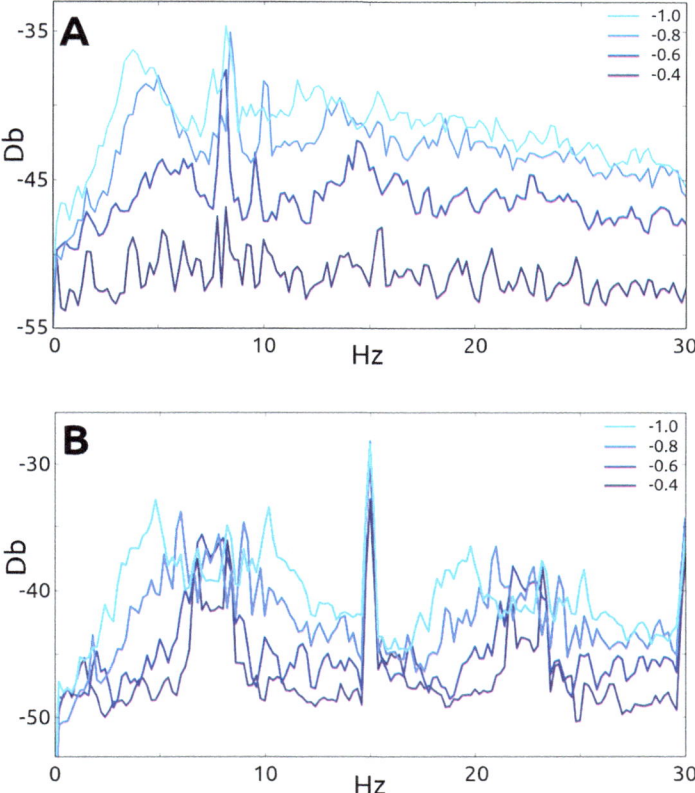

Fig. 2.2 The power spectrum of the TC neuron output with inputs from the BG in (**a**) normal state and (**b**) PD state with 15 Hz bursts, at different levels of hyper polarizing current, from −0.4 (dark) to −1.0 (light) μA

and we approximate the effect of dopaminergic depletion in the thalamus by a stronger hyperpolarizing current [4].

Figure 2.2 shows the power spectrum of the TC neurons as we increase the hyperpolarizing bias current from −0.4 to −1.0 μA. Even with the normal BG model with non-oscillatory spikes at the average rate of 65 Hz (A), as the bias current was increased, an oscillatory peak at around 7–8 Hz appeared, along with a wider band activity at a lower frequency. When the BG was in PD state with GPi oscillating at 15 Hz with the average firing rate of 75 Hz (B), in addition to a peak at 15 Hz for all level of the bias current, peaks in the 7–8 Hz range also appeared even with bias current at −0.4 and −0.6 μA.

By taking the bias current of −0.4 μA as the standard level, the results suggest that 7–8 Hz oscillation in the thalamus can be caused by either strong hyperpolarization of the TC neurons or oscillatory inhibitory input from the BG at 15 Hz range.

2.3.2   The Role of Beta-Range Oscillatory Input from the BG

Can strong inhibitory input from the BG without oscillation induce PD-like
oscillation in the thalamus? We examined this possibility by comparing the effects
of the BG input in the normal state with no synchronous oscillation and PD state
with 15 Hz burst oscillation. Figure 2.3 shows the activities of TC neurons in 3–
10 Hz range at different levels of mean firing rate (55–120 Hz) and connection
strength (0.05–0.7). With the normal BG input (A), there was little to no response
for any input levels. With the BG input in PD oscillation (B), oscillatory responses
were seen at the moderate level of input weight for a wide range of average spike
frequency. At higher weights and spike rates, the TC neurons desynchronized, and
the power in the 3–10 Hz range was decreased.

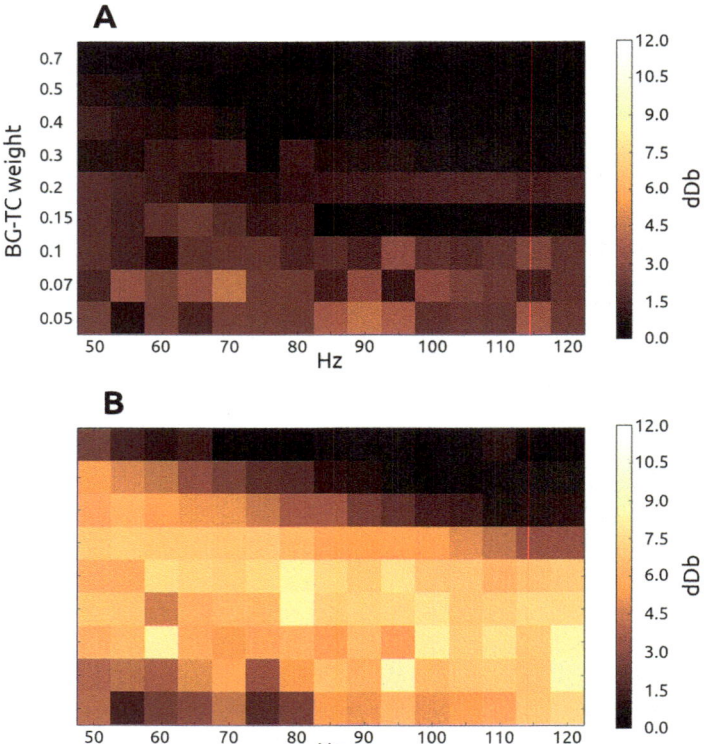

Fig. 2.3   The effects of (**a**) asynchronous and (**b**) 15 Hz synchronous bursting inputs from the GPi
on the response of TC neurons in the 3–10 Hz range. The input strengths were varied by the average
firing rate (horizontal) and input weight (vertical). Both cases have a −0.4 μA bias current

2.4 Discussion

Our spiking neural network model of the BG-TC circuit reproduced normal, asynchronous firing state and Parkinson-like state with 15 Hz beta-range oscillation in the BG and 4–8 Hz oscillation in the thalamus, which is consistent with the PD tremor frequency. Increased hyperpolarizing current in the thalamic neurons, approximating the effect of dopaminergic depletion in the thalamus, and 15 Hz synchronous oscillatory inhibitory input from the PD state of the basal ganglia can jointly trigger thalamic oscillation in the 3–10 Hz range. Stronger asynchronous inhibitory input from the BG is not sufficient to cause 3–10 Hz oscillation in the thalamus. These results imply that the effects of dopaminergic depletion in both BG and the thalamus are essential for subharmonic resonant response of the thalamic network in generating Parkinsonian tremor.

References

1. Destexhe A, Bal T, McCormick D, Sejnowski T (1996) Ionic mechanisms underlying synchronized oscillations and propagating waves in a model of ferret thalamic slices. J Neurophysiol 76(3):2049–2070
2. Djurfeldt M, Hjorth J, Eppler JM, Dudani N, Helias M, Potjans TC, Bhalla US, Diesmann M, Hellgren Kotaleski J, Ekeberg O (2010) Run-time interoperability between neuronal network simulators based on the MUSIC framework. Neuroinformatics 8(1):43–60
3. Dovzhenok A, Rubchinsky LL (2012) On the origin of tremor in Parkinson's disease. PLoS ONE 7(7):e41598
4. Edgerton J, Jaeger D (2014) Optogenetic activation of nigral inhibitory inputs to motor thalamus in the mouse reveals classic inhibition with little potential for rebound activation. Front Cell Neurosci 8:36
5. Gewaltig MO, Diesmann M (2007) Nest (neural simulation tool). Scholarpedia 2(4):1430
6. Igarashi J, Shouno O, Moren J, Yoshimoto J, Doya K (2015) A spiking neural network model of the basal ganglia-thalamo-cortical circuit toward understanding of motor symptoms of parkinson disease. Brain Neural Netw 22(3):103–111. https://doi.org/10.3902/jnns.22.103. (in Japanese)
7. Miller WC, DeLong MR (1987) Altered tonic activity of neurons in the globus pallidus and subthalamic nucleus in the primate MPTP model of Parkinsonism. Springer US, Boston, pp 415–427
8. Moren J, Igarashi J, Yoshimoto J, Doya K (2015) A full rat-scale model of the basal ganglia and thalamocortical network to reproduce parkinsonian tremor. BMC Neurosci 16(Suppl 1). https://doi.org/10.1186/1471-2202-16-s1-p64
9. Oorschot DE (1996) Total number of neurons in the neostriatal, pallidal, subthalamic, and substantia nigral nuclei of the rat basal ganglia: a stereological study using the cavalieri and optical disector methods. J Comp Neurol 366(4):580–599
10. Sarnthein J, Jeanmonod D (2007) High thalamocortical theta coherence in patients with Parkinson's disease. J Neurosci 27(1):124–131

11. Shouno O, Takeuchi J, Tsujino H (2009) A spiking neuron model of the basal ganglia circuitry that can generate behavioral variability. In: The basal ganglia IX. Springer, pp 191–200
12. Shouno O, Tachibana Y, Nambu A, Doya K (2017) Computational model of recurrent subthalamo-pallidal circuit for generation of Parkinsonian oscillations. Front Neuroanat 11:21
13. Tachibana Y, Iwamuro H, Kita H, Takada M, Nambu A (2011) Subthalamo-pallidal interactions underlying Parkinsonian neuronal oscillations in the primate basal ganglia. Eur J Neurosci 34(9):1470–1484

Chapter 3
A Neural Mass Model for Abnormal Beta-Rebound in Schizophrenia

Áine Byrne, Stephen Coombes, and Peter F. Liddle

Abstract Patients with schizophrenia demonstrate robust abnormalities of the synchronisation of beta oscillations that occur in diverse brain regions following sensory, motor or mental events. A prominent abnormality seen in primary motor cortex is a reduction in amplitude of so-called *beta-rebound*. Here a sharp decrease in neural oscillatory power in the beta band is observed during movement (MRBD) followed by an increase above baseline on movement cessation (PMBR). An understanding of how neural circuits give rise to MRBD and PMBR is clinically relevant to the pathophysiology of schizophrenia. Here we survey a very recent neural mass model for movement-induced changes in the beta rhythm and show that it is an ideal candidate for use in a clinical setting. The model arises as an exact mean-field reduction of a spiking network, has a realistic model of synaptic processing and is able to describe the dynamic changes in population synchrony that can underlie event-related desynchronisation/synchronisation for MRBD/PMBR. A lengthening of the synaptic response time to sensory drive, modelling NMDA receptor hypofunction, shows a reduction in beta-rebound consistent with that seen in schizophrenia.

Keywords Neural mass · Mean-field model · Beta-rebound · Schizophrenia

Á. Byrne (✉)
Centre for Neural Sciences, New York University, New York, NY, USA
e-mail: aine.byrne@nyu.edu

S. Coombes
Centre for Mathematical Medicine and Biology, School of Mathematical Sciences, University of Nottingham, Nottingham, UK
e-mail: stephen.coombes@nottingham.ac.uk

P. F. Liddle
Institute of Mental Health, School of Medicine, University of Nottingham, Nottingham, UK
e-mail: peter.liddle@nottingham.ac.uk

© Springer Nature Switzerland AG 2019   21
V. Cutsuridis (ed.), *Multiscale Models of Brain Disorders*, Springer Series in
Cognitive and Neural Systems 13, https://doi.org/10.1007/978-3-030-18830-6_3

3.1 Introduction

Normal cortical function depends on an intricate balance between GABAergic inhibition and glutamatergic excitation and is believed to be disrupted in neuropsychiatric disorders, such as schizophrenia [1, 2]. To build a bridge in understanding how pathophysiologies of neurons and synapses can effect cognitive processes (e.g. attention, memory, executive functions) underlying behavioural disorders (e.g. lack of goal-directed behaviour, replaying conversations out loud, catatonia), clinicians are increasingly turning to computational psychiatry [3]. This multidisciplinary approach combines insight from the clinic with mechanistic models of brain circuits that can be studied *in silico* [4]. These models range in their scale from high-dimensional models of spiking neural networks [5] to more coarse-grained phenomenological neural mass and neural field models [6], and see Breakspear [7] for a recent survey. The former are computationally expensive to simulate, though very useful for exploring ideas about neural function. For example, modelling work by Vogels and Abbott [8], using large-scale integrate-and-fire spiking neuron models, has suggested that the cognitive and behavioural deficits arising in schizophrenia may arise when a reduction in inhibition leads to a failure of sensory gating by signal-carrying pathways. This modelling study reinforces the commonly held notion of schizophrenia as a disorder of information processing. However, having a detailed microscopic dynamical description of neurons may not be necessary for building tissue activity models at the scale relevant to modern human neuroimaging studies. At this level the neural mass and field models, which maintain a good representation of synaptic dynamics, have proved to be a popular tool, especially as they can be easily realised in open-source software platforms such as The Virtual Brain [9]. The combination of neural mass models and computational platforms for the interrogation of emergent network dynamics means that one can make neuroimaging predictions as a function of distinct synaptic-level manipulations. The quality of these predictions is linked to the choice of neural mass model, which in turn should be chosen in a way to best suit the phenomenon under investigation. Here we consider the phenomenon of so-called *beta-rebound*, which is abnormal in patients with schizophrenia, and how best to develop a computational psychiatry relevant to its study.

The term 'beta-rebound' refers to the event-related synchronisation in the beta band (13–30 Hz) characteristically seen in electro- and magnetoencephalography (EEG and MEG) recordings. It can be portrayed visually using a spectrogram where it manifests as a transient increase above baseline in beta power. When recorded from motor cortex, it is often referred to as post-movement beta-rebound (PMBR) following movement cessation. This contrasts with the suppression of the beta oscillation amplitude during voluntary movement, which is referred to as movement-related beta decrease (MRBD). These modulations of the beta band are caused by changes of synchrony within a relatively localised region of motor cortex [10], with MRBD regarded as an instance of event-related desynchronisation (ERD) and PMBR of event-related synchronisation (ERS). MEG studies of patients with schizophrenia demonstrate abnormalities of PMBR [11], with a significant reduction in the trough-to-peak change in amplitude from MRBD to PMBR (although latency and duration are not significantly abnormal). Moreover, the magnitude of the abnormality predicts persistence of symptoms. Figure 3.1 illustrates the differences in the neural response between schizophrenia patients and healthy controls when performing a simple motor task. The reduction in the magnitude of the PMBR is readily observed. One should also note that the MRBD shows little to no change. Furthermore, in a salience detection task, patients with schizophrenia show a greater beta synchronisation in response to stimuli that are irrelevant than to behaviourally relevant stimuli, in motor cortex and also in the insula, a brain region engaged in detection of behavioural salience of stimuli. In contrast, healthy control participants show greater beta synchronisation in response to relevant than to irrelevant stimuli in both insula and motor cortex [12]. Thus, this abnormality of beta synchronisation is a plausible target for therapy using neuromodulatory techniques. These in turn could be assayed with the use of a realistic mechanistic model. Recent work [13] has provided such a model in the form of a *next-generation neural mass* model. This is an exact mean-field reduction of a spiking network that is capable of supporting ERD and ERS in response to a time-dependent input. Importantly it has been matched to experiments of healthy subjects exhibiting beta-rebound. Here we review the model and show that it is also capable of explaining the abnormal beta-rebound of patients with schizophrenia. This happens with an appropriate reduction in the response time to excitatory input and is consistent with the *glutamate hypothesis of schizophrenia*, which originated from the finding that phencyclidine and ketamine, which each block NMDA receptors, mimic both positive and negative symptoms

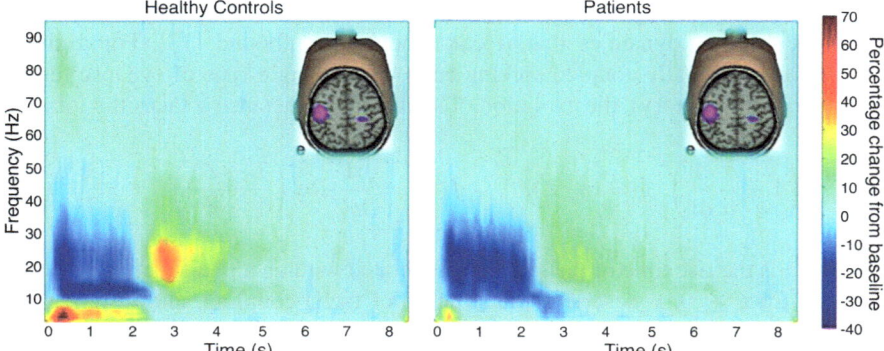

Fig. 3.1 Relative time-frequency spectrograms showing the disruption of 'beta-rebound' in schizophrenia patients. In this experiment subjects were asked to continuously tap their finger for 2 s. Movement-related beta decrease (MRBD) can be seen during the 2 s of movement, and post-movement beta-rebound (PMBR) can be seen after movement termination. Left: Healthy controls. Right: Diagnosed schizophrenia patients. Note that there is little difference in the MRBD between patients and controls, but a significant reduction in PMBR is observed for the schizophrenia patients. (Figure is reproduced and edited with permission from [11])

of disease. Thus, the model further supports the hypothesis that NMDA receptor hypofunction is a key component of the mechanism behind the pathophysiology of schizophrenia [14].

3.2 The Model

Neural mass models have a long tradition of use in describing the dynamics of neural populations. In broad terms they describe the evolution of some coarse-grained notion of neural activity, typically synaptic current or mean-membrane potential. They are formulated as coupled ordinary differential equation (ODE) models that track the dynamics of interacting excitatory and inhibitory subpopulations. Typically they are modelled as variants of the two dimensional Wilson-Cowan model [15]. With the augmentation by more realistic forms of synaptic and network interaction, they have proved especially successful in providing fits to neuroimaging data [16]. However, all neural mass models to date are essentially phenomenological, with the dynamics of state variables essentially determined by the choice of a nonlinear population firing rate function. This is often chosen to be a sigmoidal function of population activity and is a mainstay throughout computational neuroscience for rate-based modelling of neural tissue. However, the neural dynamics underlying ERD and ERS is most likely a manifestation of a *spiking* network, with enhanced ERS being linked to an increase in the coherence (synchrony) of spike trains. Thus, neural mass models in isolation are not natural candidates for modelling MRBD and PMBR. Recently a neural mass model has been *derived* as an exact mean-field model for a network of synaptically coupled spiking (quadratic integrate-and-fire) neurons and been shown to have a sufficiently rich repertoire of dynamics that it can model beta-rebound [13]. For a single population of globally coupled spiking cells with a single type of synapse (either excitatory or inhibitory), the evolution of synaptic conductance g takes the form

$$\left(1 + \frac{1}{\alpha}\frac{d}{dt}\right)^2 g = \kappa f(Z), \qquad f(Z) = \frac{1}{\pi C}\text{Re}\left(\frac{1 - Z^*}{1 + Z^*}\right). \qquad (3.1)$$

Here α^{-1} is the rise time of the synapse associated with an alpha-function response $\alpha^2 t e^{-\alpha t}$ for $t \geq 0$, κ describes the strength of (self) coupling, and C represents the membrane capacitance of the neuron. Extensions to multiple populations with different types of excitation and inhibition are straightforward and discussed in [17]. Importantly f is interpreted as a firing rate (driving the synaptic conductance) that is a function of a complex number $Z \in \mathbb{C}$ (with complex conjugate Z^*) that represents the degree of within-population synchrony. In essence, it is a Kuramoto order parameter with self-consistent dynamics determined by

$$C\frac{\mathrm{d}}{\mathrm{d}t}Z = i\frac{(Z-1)^2}{2} + \frac{(Z+1)^2}{2}[-\Delta + i(\eta_0 + A(t))] + i\frac{(Z+1)^2}{2}v_{\mathrm{syn}}g - \frac{Z^2-1}{2}g.$$

$$(3.2)$$

At the microscopic level, each neuron receives a different constant input drawn from a Lorentzian distribution with a mean η_0 and width at half maximum given by Δ (so that Δ can be used to control heterogeneity). The symbol v_{syn} denotes the reversal potential of the synapse. This is excitatory (inhibitory) if v_{syn} is positive (negative) with respect to the resting membrane potential. The function $A(t)$ represents an external signal from other brain regions. The form of the mean-field model is precisely that of a neural mass model, with the notable difference that the firing rate f is a derived quantity that is a real function of the complex Kuramoto order parameter for synchrony. This in turn is described by a complex ODE with parameters from the underlying microscopic model, so that the whole system is described by four nonautonomous ODEs. Here we take the time-dependent drive $A(t)$ to be a simple excitatory current describing transduced motor input and model it as a rectangular pulse smoothed by an alpha function with a rise time α_D^{-1} (neglecting shunts, which is reasonable for glutamatergic receptors with a large positive reversal potential). The pulse shape is explicitly given by $\Omega(t) = \Pi\Theta(t)\Theta(\tau - t)$, where Π is the amplitude of the drive, τ its duration and Θ represents a Heaviside step function.

In Fig. 3.2 we show a spectrogram of the synaptic current obtained from the model under both normal operating conditions (left) and the case of NMDA receptor hypofunction (right), modelled by a lengthening of the timescale α_D^{-1}. A reduction of beta-rebound can clearly be seen in the latter case and is consistent with that seen in MEG recordings [11].

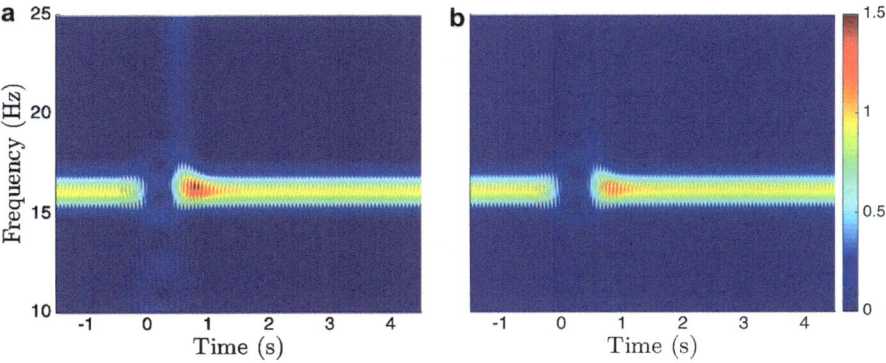

Fig. 3.2 Spectrogram of the synaptic current in response to an applied pulse at $t = 0$ s. Parameter values: $\Pi = 30\,\mu$A, $\tau = 0.4$ s, $\eta_0 = 21.5\,\mu$A, $\Delta = 0.5\,\mu$A, $v_{\mathrm{syn}} = -10$ mV, $\kappa = 0.1$, $\alpha^{-1} = 33$ ms, and $C = 30\,\mu$F. (**a**) Normal operating conditions (health) with a rise-time for the excitatory input synapse given by $\alpha_D^{-1} = 10$ ms. (**b**) NMDA receptor hypofunction (schizophrenia) with $\alpha_D^{-1} = 24$ ms. Here one can clearly see a reduction in beta-rebound compared to the model of a healthy patient

3.3 Discussion

In this chapter we have surveyed a multi-scale neural model that can capture the phenomenon of beta-rebound in healthy patients and its reduction in patients with schizophrenia. The model is multi-scale in the sense that it can be derived from an underlying spiking network in the limit of a large number of neurons as a low-dimensional set of four nonautonomous ODEs. This would rise to 12 equations should populations built from both excitatory and inhibitory neurons be considered (two ODEs for each of four synaptic populations and two complex ODEs to describe synchrony in two distinct neural populations), though is still parsimonious compared to large-scale spiking network models. Moreover, this parsimonious model is easily extended to the network level, by incorporating connectome data, and is ideally suited for computationally affordable whole brain studies. This may help in the design of optimum neuromodulatory therapeutic protocols. For example, transcranial magnetic stimulation (TMS) [18], which delivers electromagnetic stimulation to local regions of the cerebral cortex, is very effective in a minority of cases of treatment-resistant depression and also for treatment of symptoms such as hallucination in schizophrenia. However, the therapeutic effects vary greatly between cases, necessitating large clinical trials to demonstrate efficacy. The use of appropriate computational models would allow an optimisation process for therapeutic efficacy (say relating to TMS pulse sequence, location and duration of administration stimulation, etc.) to progress with a reduction in the scale of expensive clinical trials. Indeed, mathematical and computational modelling, building on the neural mass model presented here, offers the prospect of estimating the effects of administration of stimulation: (1) at the site of stimulation, (2) at remote network sites connected to the site of stimulation, (3) and, ultimately, of eventual plastic changes at relevant network sites.

References

1. Marin O (2012) Interneuron dysfunction in psychiatric disorders. Nat Rev Neurosci 13:107–120
2. Lam NH, Borduqui T, Hallak J, Roque AC, Anticevic A, Krystal JH, Wang X-J, Murray JD (2017) Effects of altered excitation-inhibition balance on decision making in a cortical circuit model. bioRxiv
3. Anticevic A, Lisman J (2017) How can global alteration of excitation/inhibition balance lead to the local dysfunctions that underlie schizophrenia? Biol Psychiatry 81:818–820
4. Lytton WW, Arle J, Bobashev G, Ji S, Klassen TL, Marmarelis VZ, Schwaber J, Sherif MA, Sanger TD (2017) Multiscale modeling in the clinic: diseases of the brain and nervous system. Brain Inform 4(4):219–230
5. Izhikevich EM, Edelman GM (2008) Large-scale model of mammalian thalamocortical systems. Proc Natl Acad Sci 105:3593–3598
6. Coombes S, Beim Graben P, Potthast R, Wright J (eds) (2014) Neural fields: theory and applications. Springer, Berlin/Heidelberg
7. Breakspear M (2017) Dynamic models of large-scale brain activity. Nat Neurosci 20:340–352

8. Vogels TP, Abbott LF (2009) Gating multiple signals through detailed balance of excitation and inhibition in spiking networks. Nat Neurosci 12:483–491
9. Sanz-Leon P, Knock SA, Spiegler A, Jirsa VK (2015) Mathematical framework for large-scale brain network modeling in the virtual brain. NeuroImage 111:385–430
10. Stancák A, Pfurtscheller G (1995) Desynchronization and recovery of beta rhythms during brisk and slow self-paced finger movements in man. Neurosci Lett 196:21–24
11. Robson SE, Brookes MJ, Hall EL, Palaniyappan L, Kumar J, Skelton M, Christodoulou NG, Qureshic A, Jan F, Liddle EB, Katshu MZ, Liddle PF, Morris PG (2016) Abnormal visuomotor processing in schizophrenia. NeuroImage: Clin 12(Supplement C):869–878
12. Liddle EB, Price D, Palaniyappan L, Brookes MJ, Robson SE, Hall EL, Morris PG, Liddle PF (2016) Abnormal salience signaling in schizophrenia: the role of integrative beta oscillations. Hum Brain Mapp 37:1361–1374
13. Byrne Á, Brookes MJ, Coombes S (2017) A mean field model for movement induced changes in the beta rhythm. J Comput Neurosci 43:143–158
14. Olney JW, Newcomer JW, Farber NB (1999) NMDA receptor hypofunction model of schizophrenia. J Psychiatric Res 33:523–533
15. Wilson HR, Cowan JD (1972) Excitatory and inhibitory interactions in localized populations of model neurons. Biophys J 12:1–24
16. Valdes-Sosa P, Sanchez-Bornot JM, Sotero RC, Iturria-Medina Y, Aleman-Gomez Y, Bosch-Bayard J, Carbonell F, Ozaki T (2009) Model driven EEG/fMRI fusion of brain oscillations. Hum Brain Mapp 30:2701–21
17. Coombes S, Byrne Á (2018) Next generation neural mass models. In: Torcini A, Corinto F (eds) Lecture notes in nonlinear dynamics in computational neuroscience: from physics and biology to ICT. Springer, Cham
18. Rogasch NC, Zafiris ZJ, Fitzgerald PB (2014) Cortical inhibition, excitation, and connectivity in schizophrenia: a review of insights from transcranial magnetic stimulation. Schizophr Bull 40:685–696

Chapter 4
Basal Ganglio-thalamo-cortico-spino-muscular Model of Parkinson's Disease Bradykinesia

Vassilis Cutsuridis

Abstract Bradykinesia is the cardinal symptom of Parkinson's disease (PD) related to slowness of movement. The causes of PD bradykinesia are not known largely, because there are multiple brain areas and pathways involved from the neuronal degeneration site (dopamine (DA) neurons in substantia nigra pars compacta (SNc) and ventral tegmental area (VTA)) to the muscles. A neurocomputational model of basal ganglio-thalamo-cortico-spino-muscular dynamics with dopamine of PD bradykinesia is presented as a unified theoretical framework capable of producing a wealth of neuronal, electromyographic, and behavioral movement empirical findings reported in parkinsonian human and animal brain studies. The model attempts to uncover how information is processed in the affected brain areas, what role does DA play, and what are the biophysical mechanisms giving rise to the observed slowness of movement in PD bradykinesia.

Keywords Parkinson's disease · Slowness of movement · Bradykinesia · Akinesia · Computer model · Dopamine · Basal ganglia · Motor cortex · Spinal cord · Triphasic pattern of muscle activation

4.1 Introduction

Bradykinesia is the hallmark and most disabling symptom of PD. Early in the disease, the most notable manifestation of bradykinesia is difficulty with walking, speaking, or getting into and out of chairs [24]. Individuals fail to swing an arm during walking or lack facial expression [1, 24, 33]. Later in life, bradykinesia

V. Cutsuridis (✉)
School of Computer Science, University of Lincoln, Lincoln, UK
e-mail: vcutsuridis@lincoln.ac.uk

© Springer Nature Switzerland AG 2019
V. Cutsuridis (ed.), *Multiscale Models of Brain Disorders*, Springer Series in Cognitive and Neural Systems 13, https://doi.org/10.1007/978-3-030-18830-6_4

affects all movements and, at its worst, results in a complete inability to move. Patients require intense concentration to overcome the apparent inertia of the limbs that exists for the simplest motor tasks. Movement initiation is particularly impaired when unnatural or novel movements are attempted [9] or when combining several movements concurrently [4, 29].

The causes of bradykinesia are not known, in part because there are multiple pathways from the sites of neuronal degeneration to the muscles. Figure 4.1 shows three of the most important pathways: (1) the pathway from SNc and VTA to the striatum and from the striatum to the substantia nigra pars reticulata (SNr) and the globus pallidus internal segment (GPi) and from there to the thalamus and the frontal cortex, (2) the pathway from SNc and VTA to the striatum and from the striatum to the SNr and the GPi and from there to the brainstem, and (3) the pathway from the SNc/VTA to cortical areas such as the supplementary motor area (SMA), the parietal cortex, and the primary motor cortex (M1), and from there to the spinal cord.

One of the popular views is that cortical motor centers are inadequately activated by excitatory circuits passing through the basal ganglia (BG) [2]. As a result, inadequate facilitation is provided to motor neuron pools, and hence movements

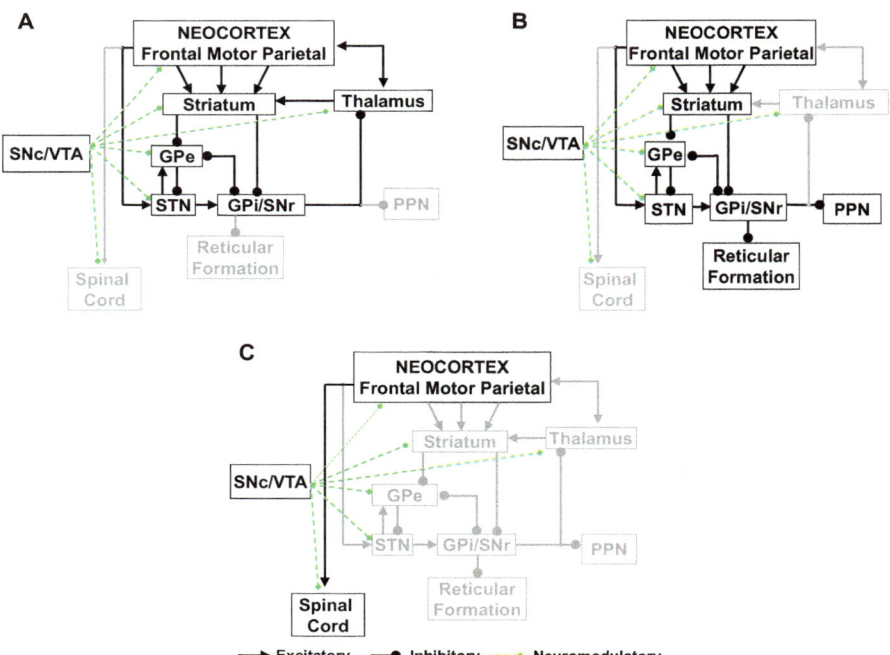

Fig. 4.1 Brain circuits implicated in PD bradykinesia

are small and weak [2]. The implication of this view is that cells in the cortex and spinal cord are functionally normally. This paper will show otherwise.

The paper's view is that disruptions of the BG output and of the SNc's DA input to frontal and parietal cortices and spinal cord are responsible for delayed movement initiation. Elimination of DA modulation from the SNc disrupts, via several pathways, the buildup of the pattern of movement-related responses in the primary motor and parietal cortex and results in a loss of directional specificity of reciprocal and bidirectional cells in the motor cortex as well as in a reduction in their activities and their rates of change. These changes result in delays in recruiting the appropriate level of muscle force sufficiently fast and in an inappropriate scaling of the dynamic muscle force to the movement parameters. A repetitive triphasic pattern of muscle activation is sometimes needed to complete the movement. All of these result in an increase of mean reaction time and a slowness of movement (i.e., bradykinesia).

4.2   Empirical Signatures of PD Bradykinesia

PD bradykinesia has been linked with the degeneration of DA neurons in SNc and VTA. Bradykinesia manifests only when 80–90% of DA neurons die. All motor cortical and subcortical areas are innervated by SNc and VTA DA neurons [6, 23, 35]. The degeneration of DA neurons leads to a number of changes relevant to bradykinesia in the neuronal, electromyographic (EMG), and movement parameters reported in parkinsonian human and animal brains:

- Reduction of peak neuronal activity and rate of development of neuronal discharge in the primary motor cortex and premotor area [26, 32].
- Abnormal oscillatory GP (external and internal) neuronal responses [31].
- Disinhibition of reciprocally tuned cells [22]. Reciprocally tuned cells are cells that discharge maximally in one movement direction but pause their activities in the opposite direction.
- Significant increase in mean duration of neuronal discharge in motor cortex preceding and following onset of movement [3, 22, 26].
- Multiple triphasic patterns of muscle activation [22, 27]. Triphasic pattern of muscle activation is a characteristic electromyographic (EMG) pattern character- ized by alternating bursts of agonist and antagonist muscles. The first agonist burst provides the impulsive force for the movement, whereas the antagonist activity provides the braking force to halt the limb. Sometimes a second agonist burst is needed to bring the limb to the final position. In PD patients, multiple such patterns are observed in order for the subjects to complete the movement.
- Reduction in the rate of development and peak amplitude of the first agonist burst of EMG activity [5, 10, 22, 25, 27, 32].

- Co-contraction of muscle activation [3]. In PD patients, the alternating agonist-antagonist-agonist muscle activation is disrupted resulting in the coactivation of opponent muscle groups.
- Increases in electromechanical delay time (time between the onset of modification of agonist EMG activity and the onset of movement) [3, 21, 22].
- Asymmetric increase in acceleration (time from movement onset to peak velocity) and deceleration (time from peak velocity till end of movement) times of a movement.
- Decrease in the peak value of the velocity trace [3, 8, 21, 22, 25, 30, 34].
- Significant increases in movement time [3, 21, 22, 30, 32, 34].

4.3   Basal Ganglio-thalamo-cortico-spino-muscular Model of PD Bradykinesia

Figure 4.2 depicts the basal ganglio-thalamo-cortico-spino-muscular model with dopamine of PD bradykinesia dynamics. The mathematical formalism of the model has been detailed in [11, 16, 20]. The model is composed of three modules coupled together: (1) the basal ganglio-thalamic module, (2) the cortical module, and (3) the spino-muscular module. All modules and their components are modulated by DA. The basal ganglio-thalamic module generates a scalable voluntary GO signal that gates volitional-sensitive velocity motor commands in the cortical module, which activate the lower spinal centers in the spino-muscular module. In the cortical module, an arm movement difference vector (DV) is computed in cortical parietal area 5 from a comparison of a target position vector (TPV) with a representation of the current position called perceived position vector (PPV). The DV signal then projects to area 4 (primary motor cortex), where a desired velocity vector (DVV) and a non-specific co-contractive signal (P) [28] are formed. The DVV and P signals correspond to two partly independent neuronal systems within the motor cortex. DVV represents the activity of reciprocal neurons [22] and is organized for the reciprocal activation of antagonist muscles. P represents the activity of bidirectional neurons (i.e., neurons whose activity decreases or increases for both directions of movement [22]) and is organized for the co-contraction of antagonist muscles. Whereas the reciprocal pattern of muscle activation serves to move the joint from an initial to a final position, the antagonist co-contraction serves to increase the apparent mechanical stiffness of the joint, thus fixing its posture or stabilizing its course of movement in the presence of external force perturbations [7, 28]. The spino-muscular module is an opponent-processing muscle control model of how spinal circuits afford independent voluntary control of joint stiffness and joint position. It incorporates second-order dynamics, which play a large role in realistic limb movements.

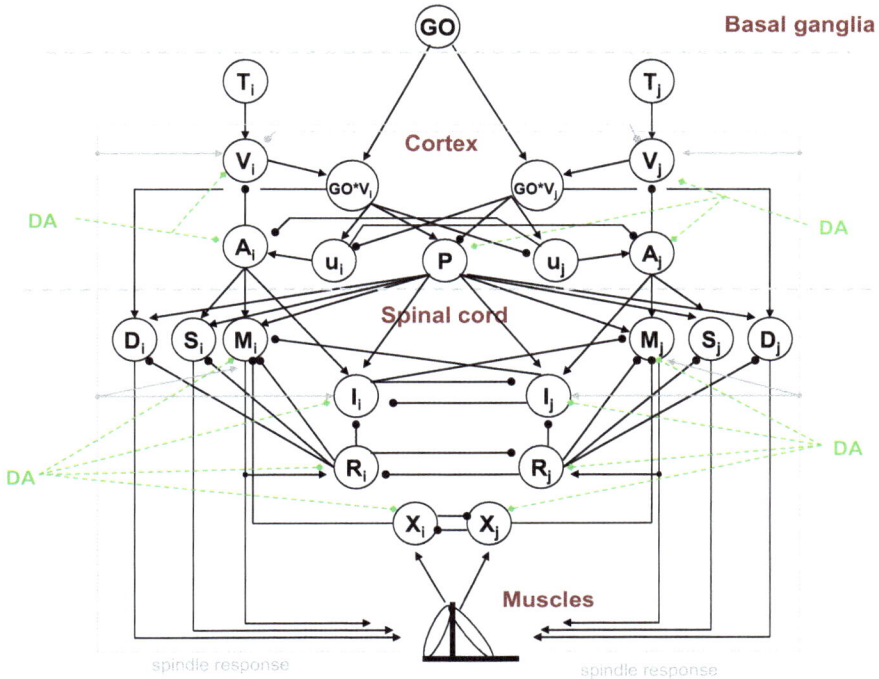

Fig. 4.2 Neural architecture of the dopamine modulated basal ganglio-thalamo-cortico-spino-muscular model. *Top*: basal ganglia module (GO signal representing the opening of the thalamo-cortical gate via inhibition of GPi response). *Middle*: cortical module for trajectory formation. *Bottom*: opponent-processing spino-muscular module for agonist-antagonist-agonist muscle activation. Arrow black lines, excitatory projections; solid dot black lines, inhibitory projections; diamond-dotted green lines, dopamine modulation; solid arrow gray lines, excitatory feedback pathways from muscle spindles. Solid dot gray lines: inhibitory feedback pathways from muscle spindles. $\underline{GO}$ globus pallidus internal segment (GPi) output signal, P bidirectional co-contractive signal, T target position command, V difference vector (DV) activity, u desired velocity vector (DVV) activity, A current perceived position vector (PPV) activity, M alpha motoneuronal activity, R Renshaw cell activity, X spinal type-b inhibitory interneuronal activity, I spinal type-a inhibitory interneuronal activity, S static gMN activity, D dynamic gMN activity, i and j antagonist cell pair

4.4   Results

The model can account for all empirical signatures of PD bradykinesia as they have been described in the previous section and reported in previous publications of the author [11–20]. To assist the readers of this paper, a subset of these simulation results are reported here. Reduction of DA in cortical and subcortical motor areas

34 V. Cutsuridis

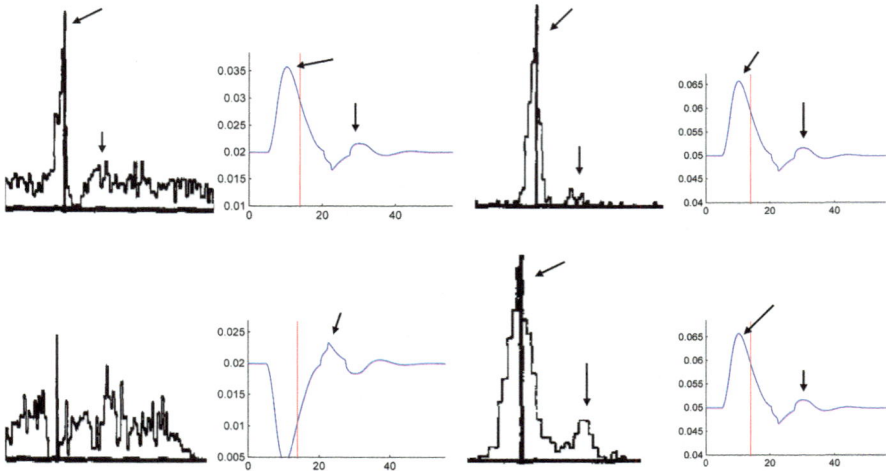

Fig. 4.3 Comparison of peristimulus time histograms (PSTH) of reciprocally organized neurons (column 1; reproduced with permission from Doudet et al. [22, Fig. 4.4a, p. 182], Copyright Springer-Verlag) in area 4, simulated area's 4 reciprocally organized phasic (DVV) cell activities (column 2), PSTH of area's 4 bidirectional neurons (column 3; reproduced with permission from [22, Fig. 4.4a, p. 182], Copyright Springer-Verlag), and simulated area's 4 co-contractive (P) cells activities (column 4) for a flexion (row 1) and extension (row 2) movements in normal monkey. The *vertical bars* indicate the onset of movement. Note a clear triphasic AG1-ANT1-AG2 pattern marked with *arrows* is evident in PSTH of reciprocally and bidirectionally organized neurons. The same triphasic pattern is evident in simulated DVV cell activities. The second peak in simulated activities marked with an arrow arises from the spindle feedback input to area's 5 DV activity

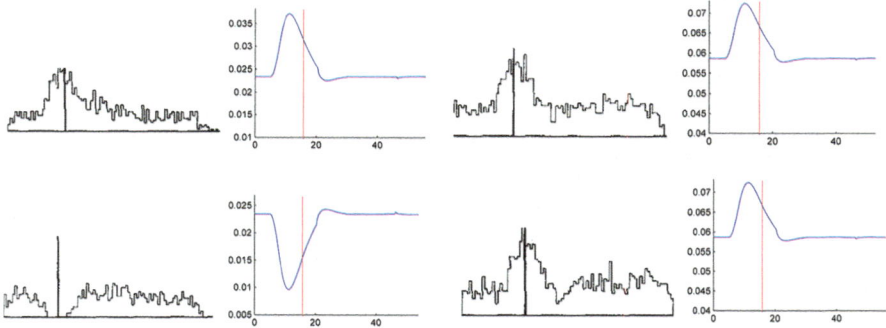

Fig. 4.4 Comparison of PSTH of reciprocally organized neurons (column 1; reproduced with permission from [22, Fig. 4.4a, p. 182], Copyright Springer-Verlag) in area 4, simulated area's 4 reciprocally organized phasic (DVV) cell activities (column 2), PSTH of area's 4 bidirectional neurons (column 3; reproduced with permission from [22], Fig. 4.4a, p. 182, Copyright Springer-Verlag), and simulated area's 4 co-contractive (P) cells activities (column 4) for a flexion (**a** and **c**) and extension (**b** and **d**) movements in MPTP-treated monkey. The *vertical bars* indicate the onset of movement. Note that the triphasic pattern is disrupted: Peak AG1 and AG2 bursts have decreased, and ANT pause is shortened

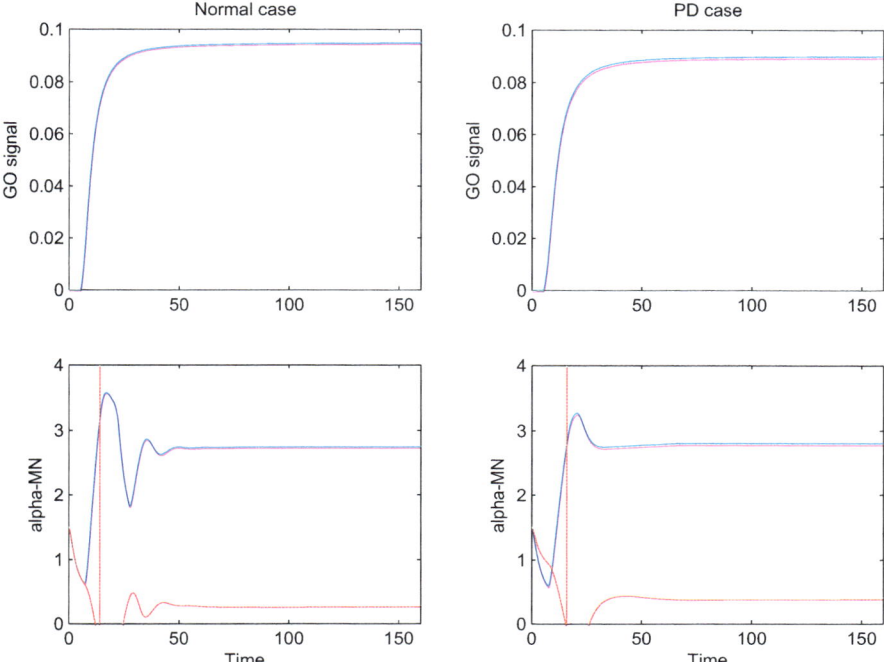

Fig. 4.5 Comparison of simulated GO signals (row 1) and α-MN activities (row 2) in normal (column 1) and dopamine-depleted (column 2) conditions. (Row 2) *Blue solid curve*, agonist α-MN activity; *Red-dashed curve*, antagonist α-MN activity. Note in PD (DA-depleted) case, the triphasic pattern is disrupted, and it is replaced by a biphasic pattern of muscle activation. Also, the peaks of agonist and antagonist bursts are decreased

disrupts, via several pathways, the rate of development and peak neuronal activity of primary motor cortical cells (reciprocal and bidirection neurons) (see Figs. 4.3 and 4.4 for comparison). A clear triphasic AG1-ANT1-AG2 pattern marked with *arrows* which is evident in control case PSTH of reciprocally and bidirectionally organized neurons (Fig. 4.3) disappears in the dopamine-depleted case (Fig. 4.4). The same triphasic pattern is evident in simulated control DVV cell activities (Fig. 4.3) that disappears in the DA-depleted case (Fig. 4.4).

These changes lead in delays in recruiting the appropriate level of muscle force sufficiently fast and in a reduction of the peak muscle force required to complete the movement (see Fig. 4.5).

Repetitive and sometimes co-contractive patterns of muscle activation are needed to complete the movement (see Fig. 4.6).

These disruptions result in an abnormal slowness of movement (see Fig. 4.7).

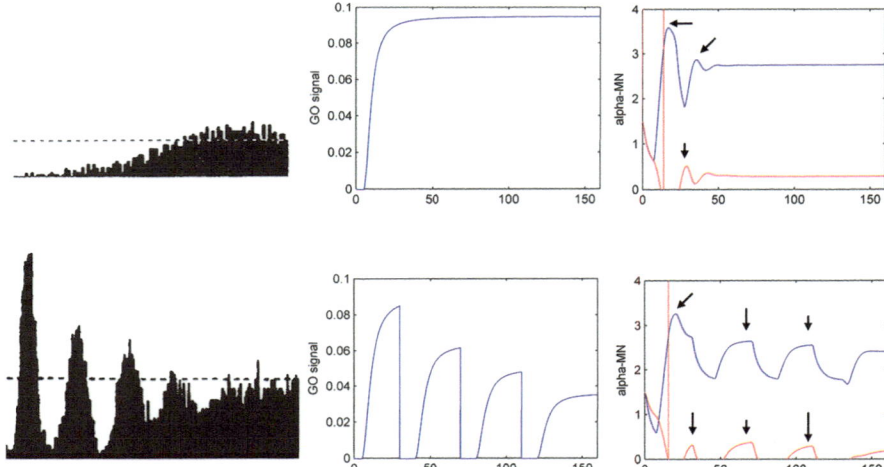

Fig. 4.6 Comparison of the experimental GPi PSTH (column 1), GO signals (column 2), and α-MN activities (column 3) in normal (row 1) and dopamine-depleted (row 2) conditions. (*Column 3, rows 1 and 2*) *Blue-colored solid curve*, agonist α-MN unit; *Red-colored dashed curve*, antagonist α-MN unit. Note in dopamine-depleted case, the α-MN activity is disrupted and replaced by repetitive and co-contractive agonist-antagonist bursts (*row 2, column 3*). (*Column 1, row 1*) GPi PSTH in intact monkey reproduced with permission from Tremblay et al. [49, Fig. 4.4, p. 6], Copyright Elsevier. (*Column1, row 2*) GPi PSTH in MPTP monkey reproduced with permission from [49, Fig. 4.2, p. 23], Copyright Elsevier

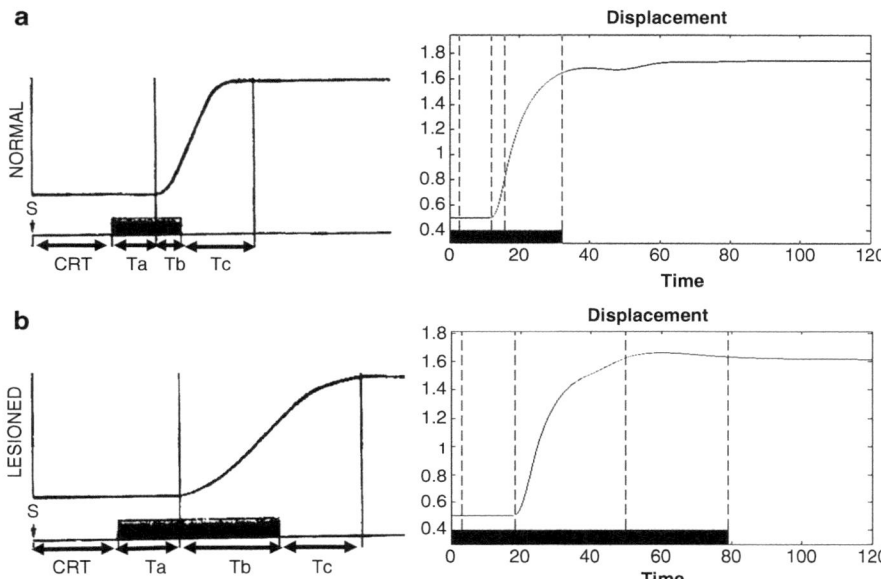

Fig. 4.7 Comparison of experimentally obtained (column 1; adapted from [26], Fig. 4.5, p. 189) and simulated (column 2) forearm displacement (position) in normal (**a**) and Parkinson's disease (**b**) conditions. *Shaded area*: representation of neuronal change related to movement. In **a** and **b**, (*column 1*) the vertical bar indicates the onset of forearm displacement; *S* auditory cue, *CRT* mean value of cellular reaction time, T_A changes of neuronal activity preceding onset of movement (OM), T_B changes of neuronal activity following OM, T_C changes of neuronal activity from T_B till movement end (ME). (*column 2*) The time interval between vertical-dashed lines starting from left to right indicates CRT, T_A, T_B, and T_C, respectively

References

1. Abbs JH, Hartman DE, Vishwanat B (1987) Orofacial motor control impairment in Parkinson's disease. Neurology 37:394–398
2. Albin RL, Young AB, Penney JB (1989) The functional anatomy of basal ganglia disorders. TINS 12:366–375
3. Benazzouz A, Gross C, Dupont J, Bioulac B (1992) MPTP induced hemiparkinsonism in monkeys: behavioral, mechanographic, electromyographic and immunohistochemical studies. Exp Brain Res 90:116–120
4. Benecke R, Rothwell JC, Dick JPR (1986) Performance of simultaneous movements in patients with Parkinson's disease. Brain 109:739–757
5. Berardelli A, Dick JPR, Rothwell JC, Day BL, Marsden CD (1986) Scaling of the size of the first agonist EMG burst during rapid wrist movements in patients with Parkinson's disease. J Neurol Neurosurg Psychiatry 49:1273–1279
6. Bjorklund A, Lindvall O (1984) Dopamine containing systems in the CNS. In: Bjorklund A, Hokfelt T (eds) Handbook of chemical neuroanatomy. Classical transmitters in the CNS, Part 1, vol 2. Elsevier, Amsterdam, pp 55–121

7. Bullock D, Contreras-Vidal JL (1993) How spinal neural networks reduce discrepancies between motor intention and motor realization. In: Newel K, Corcos D (eds) Variability and motor control. Human Kinetics Press, Champaign., 1993, pp 183–221

8. Camarata PJ, Parker RG, Park SK, Haines SJ, Turner DA, Chae H et al (1992) Effects of MPTP induced hemiparkinsonism on the kinematics of a two-dimensional, multi-joint arm movement in the rhesus monkey. Neuroscience 48(3):607–619

9. Connor NP, Abbs JH (1991) Task-dependent variations in parkinsonian motor impairments. Brain 114:321–332

10. Corcos DM, Chen CM, Quinn NP, McAuley J, Rothwell JC (1996) Strength in Parkinson's disease: relationship to rate of force generation and clinical status. Ann Neurol 39(1):79–88

11. Cutsuridis V (2006a) Biologically inspired neural architectures of voluntary movement in normal and disordered states of the brain. Unpublished PhD dissertation

12. Cutsuridis V (2006b) Neural model of dopaminergic control of arm movements in Parkinson's disease bradykinesia. Kolias S, Stafilopatis A, Duch W ICANN 2006: artificial neural networks. LNCS, 4131. Springer, Berlin, 583–591

13. Cutsuridis V (2007) Does reduced spinal reciprocal inhibition lead to cocontraction of antagonist motor units? A modeling study. Int J Neural Syst 17(4):319–327

14. Cutsuridis V (2010a) Neural network modeling of voluntary single joint movement organization. I. Normal conditions. In: Chaovalitwongse WA et al (eds) Computational neuroscience. Springer, New York, pp 181–192

15. Cutsuridis V (2010b) Neural network modeling of voluntary single joint movement organization. II. Parkinson's disease. In: Chaovalitwongse WA, Pardalos P, Xanthopoulos P (eds) Computational neuroscience. Springer, New York, pp 193–212

16. Cutsuridis V (2011) Origins of a repetitive and co-contractive pattern of muscle activation in Parkinson's disease. Neural Netw 24:592–601

17. Cutsuridis V (2013a) Bradykinesia models of Parkinson's disease. Scholarpedia 8(9):30937

18. Cutsuridis V (2013b) Bradykinesia models. In: Jaeger D, Jung R (eds) Encyclopedia of computational neuroscience. Springer, New York

19. Cutsuridis V (2018) Bradykinesia models. In: Jaeger D, Jung R (eds) Encyclopedia of computational neuroscience, 2nd edn. Springer, New York

20. Cutsuridis V, Perantonis S (2006) A neural model of Parkinson's disease bradykinesia. Neural Netw 19(4):354–374

21. Doudet DJ, Gross C, Lebrun-Grandie P, Bioulac B (1985) MPTP primate model of Parkinson's disease: a mechanographic and electromyographic study. Brain Res 335:194–199

22. Doudet DJ, Gross C, Arluison M, Bioulac B (1990) Modifications of precentral cortex discharge and EMG activity in monkeys with MPTP induced lesions of DA nigral lesions. Exp Brain Res 80:177–188

23. Gerfen CR, Engber TM, Mahan LC, Susel Z, Chase TN, Monsma FJ Jr, Sibley DR (1990) D1 and D2 dopamine receptor-regulated gene expression of striatonigral and striatopallidal neurons. Science 250:1429–1432

24. Gibberd FB (1986) The management of Parkinson's disease. Practitioner 230:139–146

25. Godaux E, Koulischer D, Jacquy J (1992) Parkinsonian bradykinesia is due to depression in the rate of rise of muscle activity. Ann Neurol 31(1):93–100

26. Gross C, Feger J, Seal J, Haramburu P, Bioulac B (1983) Neuronal activity of area 4 and movement parameters recorded in trained monkeys after unilateral lesion of the substantia nigra. Exp Brain Res 7:181–193

27. Hallett M, Khoshbin S (1980) A physiological mechanism of bradykinesia. Brain 103:301–314

28. Humphrey DR, Reed DJ (1983) Separate cortical systems for control of joint movement and joint stiffness: reciprocal activation and coactivation of antagonist muscles. In: Desmedt JE (ed) Motor control mechanisms in health and disease. Raven Press, New York. 1983

29. Lazarus JC, Stelmach GE (1992) Inter-limb coordination in Parkinson's disease. Mov Disord 7:159–170

30. Rand MK, Stelmach GE, Bloedel JR (2000) Movement accuracy constraints in Parkinson's disease patients. Neuropsychologia 38:203–212
31. Tremblay L, Filion M, Bedard PJ (1989) Responses of pallidal neurons to striatal stimulation in monkeys with MPTP-induced parkinsonism. Brain Res 498(1):17–33
32. Watts RL, Mandir AS (1992) The role of motor cortex in the pathophysiology of voluntary movement deficits associated with parkinsonism. Neurol Clin 10(2):451–469
33. Weiner WJ, Singer C (1989) Parkinson's disease and non-pharmacologic treatment programs. J Am Geriatr Soc 37:359–363
34. Weiss P, Stelmach GE, Adler CH, Waterman C (1996) Parkinsonian arm movements as altered by task difficulty. Parkinsonism Relat Disord 2(4):215–223
35. Williams SM, Goldman-Rakic PS (1998) Widespread origin of the primate mesofrontal dopamine system. Cereb Cortex 8:321–345

Chapter 5
Network Models of the Basal Ganglia in Parkinson's Disease: Advances in Deep Brain Stimulation Through Model-Based Optimization

Karthik Kumaravelu and Warren M. Grill

Abstract Parkinson's disease (PD) is a movement disorder resulting from degeneration of dopaminergic neurons in the substantia nigra pars compacta. Electrical stimulation of the sub-cortical regions of the brain (basal ganglia – BG), also known as deep brain stimulation (DBS), is an effective therapy for the motor symptoms of PD. However, despite clear clinical benefits, the therapeutic mechanisms of DBS are not fully understood. Computational models of the BG play a vital role in investigation of the neural basis of PD and determining the therapeutic mechanisms of DBS. We review several conductance-based computational models of the BG published in the literature. First, we explain the different circuits within the BG network associated with movement control. Second, we provide insights gained from different computational models of the BG on the neural basis of PD and therapeutic mechanisms of DBS. Third, we discuss the functionality of these models to optimize DBS parameters. Finally, we present various opportunities available to optimize further DBS therapy by laying out the critical elements lacking in existing models.

Keywords Basal ganglia network model · Parkinson's disease · Deep brain stimulation · Model-based optimization · Subcortical lesion · Subthalamic nucleus · Movement disorders · Computational neuroscience · Globus pallidus

K. Kumaravelu
Department of Biomedical Engineering, Duke University, Durham, NC, USA

W. M. Grill (✉)
Department of Biomedical Engineering, Duke University, Durham, NC, USA

Department of Electrical and Computer Engineering, Duke University, Durham, NC, USA

Department of Neurobiology, Duke University, Durham, NC, USA

Department of Neurosurgery, Duke University, Durham, NC, USA
e-mail: warren.grill@duke.edu

© Springer Nature Switzerland AG 2019   41
V. Cutsuridis (ed.), *Multiscale Models of Brain Disorders*, Springer Series in
Cognitive and Neural Systems 13, https://doi.org/10.1007/978-3-030-18830-6_5

5.1 Introduction

Parkinson's disease (PD) is a movement disorder resulting from degeneration of dopaminergic neurons in the substantia nigra pars compacta. Functional neurosurgical interventions such as subcortical lesions and deep brain stimulation (DBS) are both effective in suppressing the motor symptoms of PD. Computational models of the basal ganglia (BG) provide a platform to investigate the neural basis of parkinsonian symptoms and to study and optimize the therapeutic effects of DBS. In this chapter, we provide a critical review of several conductance-based network models of the BG available in the literature. First, we describe the key components of the BG network and its effects on the thalamus. Second, we review different network models of the BG and their potential to explain the neural basis of PD symptoms and the therapeutic mechanisms of DBS. Third, we provide an overview of the use of computational models of the BG network to optimize the parameters of DBS. The two major issues with current DBS technology are stimulation-evoked side effects [82] and the requirement for frequent replacement of the implanted pulse generator (IPG) due to depleted batteries [14]. The process of selection of DBS parameters is challenging due to the large number of degrees of freedom and complexity of response to different parameters [59]. Model-based optimization provides a systematic framework for the selection of DBS parameters. Finally, we consider the critical aspects lacking in existing network models of BG that might enable better optimization of DBS therapy.

5.2 Architecture of the Basal Ganglia Circuit

The basal ganglia (BG) is a group of subcortical nuclei in the brain comprised of the substantia nigra pars compacta (SNc), striatum (Str), subthalamic nucleus (STN), globus pallidus externa (GPe), globus pallidus interna (GPi), and substania nigra pars reticulata (SNr) (Fig. 5.1). The Str and STN serve as the input nuclei of the BG and receive dense excitatory projections from the cortex (CTX) [34, 35]. The GPi and SNr are the two primary output nuclei of the BG and send inhibitory projections to the thalamus (TH) [15]. Medium spiny neurons of the Str are modulated by the neurotransmitter dopamine via D1 and D2 receptors [15], and the GPe and GPi/SNr receive inhibitory projections from the D2- and D1-modulated Str neurons, respectively [15]. The STN sends excitatory projections to both GPe and GPi/SNr [15], and the GABAergic GPe neurons make reciprocal projections back to the STN (resulting in a closed loop STN-GPe network) as well as to the GPi/SNr [85]. The Str neurons modulate GPi/SNr via two pathways: (1) Str (D1) → GPi known as the direct pathway and (2) Str (D2)→GPe→GPi referred to as the indirect pathway. Furthermore, the CTX exerts influence on the GPi via direct projections to the STN known as the hyperdirect pathway. Motor programs from the CTX are differentially modulated at the Str by dopaminergic

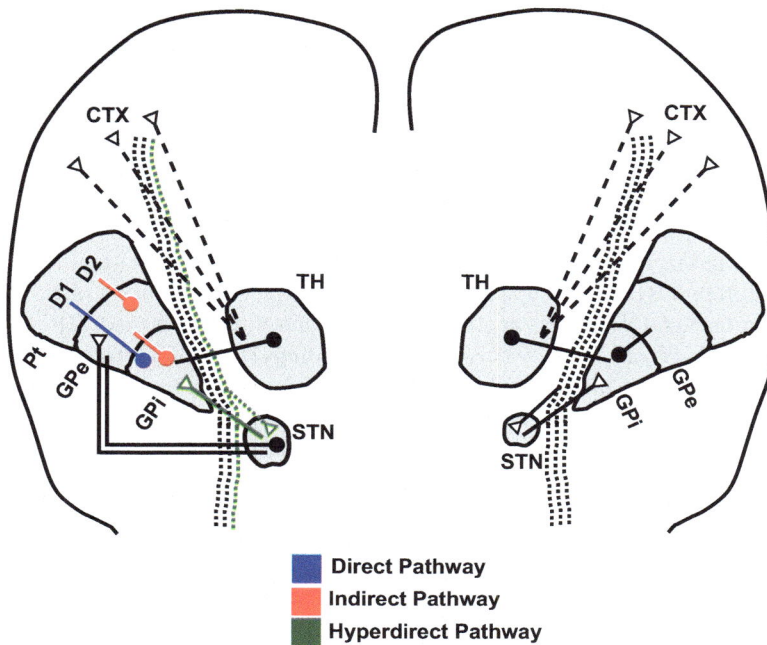

Fig. 5.1 Architecture of basal ganglia (BG) network. The BG circuit comprises of three major pathways – (1) D1-Str→GPi known as the direct pathway shown in blue, (2) D2-Str→GPe→GPi known as the indirect pathway shown in red and, (3) CTX→STN→GPi known as the hyperdirect pathway shown in green. Any comprehensive computational model of the BG circuit should capture these three pathways

neurons projecting from the SNc. The classical rate model of PD demonstrates the functional role of dopamine on the BG pathways [4]. In accordance with the rate model, activation of the direct pathway via the D1 receptors results in inhibition of GPi and subsequent disinhibition of the TH. Indirect pathway activation through the D2 receptors results in excitation of GPi and subsequent inhibition of the TH. However, recent findings challenging the rate model suggest the direct and indirect pathways to be not functionally distinct with both the pathways being interlinked at the level of Str (via interneurons) and GPe (via collaterals) [20, 25].

5.3 Neural Basis of Motor Symptoms of PD Explained Using Network Models of the Basal Ganglia

Parkinson's disease (PD) involves the degeneration of dopaminergic neurons in the SNc [3]. The primary motor symptoms of PD are akinesia/bradykinesia, rest tremor, rigidity, postural instability, and gait disorder [53]. Initial attempts were made to explain the underlying changes in neural activity resulting in PD motor symptoms using the classical rate model of the BG. According to this model [4], SNc dopamine depletion results in an imbalance between the direct and indirect BG pathways resulting in decreased activation of the direct pathway and increased activation of the indirect pathway. Increased activation of the indirect pathway leads to a decrease in GPe firing rate and a subsequent increase in the STN and GPi firing rates. The firing rate of GPi is further increased by the decreased activation of the direct pathway. Therefore, a hyperactive GPi during PD provides a greater inhibition to the TH, which results in bradykinesia/akinesia. Single-unit recordings across different BG nuclei support the classical rate model. The firing rates of STN, GPi neurons in 6-OHDA rats are higher than in control, while those of GPe neurons are lower [44, 49, 68, 69, 76, 93]. Similarly, recordings from MPTP-treated monkeys show an increase in the STN and GPi firing rates but a decrease in the GPe firing rate post MPTP treatment [11, 13, 33, 47, 87, 97, 98].

The rate model excluded a potential role of the hyperdirect pathway in the manifestation of PD symptoms. The functional significance of the hyperdirect pathway in the normal execution of movements can be explained using the center-surround theory proposed by Nambu and colleagues [75]. When a movement is initiated, the motor cortex exerts a strong excitatory influence on the output nucleus of the basal ganglia (GPi) via the hyperdirect pathway to the STN [74]. The excitation of GPi and subsequent inhibition of TH is thought to enable selection of one motor program through the negation of competing programs. The activation of the hyperdirect pathway is followed by the activation of the direct pathway which results in inhibition of GPi and subsequent disinhibition of TH. This command enables proper selection of the desired motor program. Finally, the activation of the indirect pathway results in increased GPi activity and TH inhibition. This aids in the suppression of unwanted motor programs and further enables transmission of the selected motor command.

The hyperdirect pathway is known to become abnormal during PD condition [9, 24, 72], and therefore, a more accurate network model of BG should account for the critical role played by the hyperdirect pathway during normal execution of movements and subsequent manifestation of PD symptoms following its alterations in PD state. Humphries and colleagues developed a high-level network model of the BG circuit incorporating the center-surround action selection theory proposed by Nambu and colleagues [52]. The model predicted that under the dopamine-depleted condition, the BG circuit failed to switch appropriately between actions initiated at the CTX. Thus, the model implicates the faulty action selection mechanism of BG for the underlying motor symptoms of PD.

5.4 Therapeutic Mechanisms of Lesion Explained Using Network Models of the Basal Ganglia

Lesion of the subcortical regions of the brain is effective in treating PD motor symptoms [39, 45, 60, 66, 71]. The therapeutic mechanism of STN lesion can be explained using the classical rate model of PD. STN lesion would reduce the activity of an already hyperactive parkinsonian GPi and result in a reduction of net inhibition to the TH and thereby reduce bradykinesia. Similarly, lesion of GPi would lead to a reduction of the overall inhibition to the TH and thereby reduce bradykinesia. Thus, the classical rate model was sufficient to explain the therapeutic mechanisms of subcortical lesion. Despite the clinical effectiveness of lesions in suppressing PD motor symptoms, the effects are irreversible and non-adjustable, and especially bilateral lesions can be associated with unacceptable side effect profiles. This necessitated the development of an alternative therapy which could suppress PD motor symptom akin to lesion as well as provide greater control in adjusting treatment [10].

5.5 Therapeutic Mechanisms of DBS Explained Using Network Models of the Basal Ganglia

Chronic high-frequency (HF) stimulation of the STN and GPi are effective at treating the motor symptoms of advanced PD, including tremor and bradykinesia [38, 56, 62], and provide outcomes that are superior to conventional medical management [27, 96]. In PD patients, STN DBS at frequencies above 100 Hz provides clinical benefits, while frequencies below 50 Hz are usually ineffective [73]. Despite the clinical effectiveness of HF STN DBS, the therapeutic mechanisms of this therapy are not entirely understood [8, 48].

Efforts were made to explain the therapeutic mechanism of STN DBS using the classical rate model of PD. Results from single-cell computational models [37, 70] and experimental studies suggest that DBS activates the afferent and efferent axons projecting to and from the stimulated nucleus. For example, STN DBS results in activation of axons projecting from the STN to GPe and GPi [43]. Similarly, GPi DBS results in activation of the GPi efferent axons to TH [6]. Therefore, both STN and GPi DBS result in increased firing of GPi neurons. However, according to the classical rate model, increasing the activity of GPi by STN DBS should lead to a *more* bradykinetic state. This prediction of the rate model is in contrast with the clinical outcomes observed during GPi and STN DBS in PD patients, where bradykinesia is reduced. Therefore, the classical rate model is not sufficient to explain the therapeutic benefits observed during STN and GPi DBS in PD. This necessitated the development of computational models of the BG network to represent the neural activity patterns in addition to firing rates seen in PD and how these are altered by therapeutically effective DBS.

Rubin and Terman (RT) developed a biophysical model of the BG network [84]. The BG neurons were modeled using single-compartment Hodgkin-Huxley (HH)-style neurons, and the network model accounted for differences in neural activity observed in healthy and PD conditions. In the PD state, model BG neurons exhibited synchronous oscillatory activity in the theta band similar to those seen in vitro [78]. The model was then used to explore the therapeutic effects of HF STN DBS. Error index, a measure characterizing the efficacy of the TH to function as a relay, was used as a model-based proxy for symptom to quantify the therapeutic effects of STN DBS. The robustness of error index as a model-based proxy for PD symptoms was quantified in a separate study [41]. GPi activities recorded in nonhuman primates under "healthy," "PD," "PD + subtherapeutic DBS," and "therapeutic DBS" were fed as inputs to the TC neurons from the RT model. The model predicted a loss in thalamic relay fidelity (i.e., an increase in error index) during input recorded under PD condition compared to healthy, and the fidelity was restored only during therapeutic but not subtherapeutic DBS. However, the RT model did not account for the frequency-dependent effects of STN DBS in suppressing PD symptoms [86]. In the RT model, STN DBS frequencies greater than 20 Hz were effective in suppressing PD symptoms, which is inconsistent with clinical observations [73]. Several studies used the RT model to optimize DBS targets [77], amplitude/pulse width [23, 32, 88, 95], multi-site stimulation [7, 40], multi-input phase-shifted patterns [2], irregular/random patterns [90], and closed-loop controllers [36, 63, 64, 89].

The STN-GPe subcircuit from the RT model [92], incorporating synaptic plasticity, has been used extensively to quantify the effects of a novel DBS protocol called coordinated reset [30, 31, 46, 65]. Coordinated reset (CR) includes phase-shifted stimulation trains delivered via multiple electrodes with the aim of desynchronizing the excessive synchronization between neurons within BG nuclei and between nuclei of the BG network that occurs in PD [12]. The model predicts spike-time-dependent plasticity between intra-STN and intra-GPe synapses which can be exploited to induce a long-lasting therapeutic response to CR stimulation compared to continuous HF stimulation. A few pilot studies have been attempted in PD patients, and MPTP-treated nonhuman primates to validate the model predicted long-lasting neural effects of the CR protocol [1, 91, 94]. However, further experimental studies are required to compare the neural effects of CR protocol with model-based predictions, and if the effects match, then the model can be used further to optimize the CR protocol [67].

Another novel DBS protocol known as delayed feedback (DF) is a closed-loop technique where high-frequency stimulation is delivered at a set delay with the amplitude of the high-frequency train being modulated based on oscillatory activity detected in the local field potential. The STN-GPe subcircuit from the RT network predicts the DF protocol to be also effective in desynchronizing the ongoing oscillatory activity [79]. An adaptive DBS (aDBS) scheme similar to the DF protocol was tested in PD patients, and aDBS was found to be more efficient in suppressing PD motor symptoms at a much lower energy compared to HF DBS [83]. These studies using the RT model revealed desynchronization of abnormal oscillatory activity as an important therapeutic mechanism of DBS.

A revised version of the RT model was developed to account for the strong frequency-dependent effects of STN DBS [86]. The model BG neurons in the So model exhibited synchronous beta oscillations in the PD state similar to those observed in bradykinesia-dominant PD patients [18, 22]. Pathological beta oscillations are strongly correlated with motor symptoms of PD especially bradykinesia and may serve as a proxy to evaluate the efficacy of DBS [57, 81]. The So version of the RT model was used to quantify the effects of novel DBS waveforms [26]. The model predicted the delayed Gaussian waveform to outperform the traditional rectangular waveform both in efficiency to elicit action potentials and in energy consumption. Further, the So version of the RT model was used in multiple studies to characterize the neural response to different temporal patterns of DBS. First, temporally random patterns of STN DBS with an average frequency of 130 Hz performed poorly in reducing bradykinesia in PD patients compared to regular 130 Hz [28]. A modified version of the RT model akin to the So model was used to quantify the underlying neural response responsible for this effect. Random STN DBS in the model failed to regularize GPi neural activity and restore thalamic relay function. Second, in a different study in PD patients, several temporally non-regular patterns of DBS were found to be more effective in suppressing bradykinesia compared to regular 185 Hz STN DBS [17]. The So version of the RT model was used to demonstrate that the non-regular patterns of STN DBS differentially modulated beta-band power in the activity of model GPi neurons with a trend similar to the measured motor response in PD patients [17]. Finally, the So version of the RT model was coupled to a genetic algorithm to optimize temporal patterns of STN DBS, and the model-based optimized pattern with an average frequency of 45 Hz was efficient in suppressing motor symptoms in PD patients similar to regular HF STN DBS [16]. These studies quantifying the effects of temporal patterns of DBS were instrumental in revealing that HF stimulation is neither necessary nor sufficient to generate effective symptom relief.

A biophysically based network model of the BG developed by Hahn and McIntyre (HM) also successfully accounted for the frequency-dependent effects of STN DBS [42]. In contrast to the So/RT model, the BG network in the HM model was optimized to match in vivo firing rates recorded from nonhuman primates. The model BG neurons in the HM model exhibited increased burst activity and synchronous beta oscillations in the PD state compared to the healthy condition. In the HM model, reduction of GPi burst activity in a stimulation frequency-dependent manner was theorized as a potential therapeutic mechanism of DBS. Subsequently, the HM model was used to obtain optimum phase response curves for STN DBS [50, 51].

Kang and Lowery extended the BG circuit by modeling the cortical neurons and their connections to STN, the hyperdirect pathway [54]. The model implicated the hyperdirect pathway as the entry point to the BG of abnormal low-frequency oscillations during PD and added support to the desynchronization theory as the basis for the therapeutic mechanism of DBS. The model was then used as a test bed to evaluate the performance of several closed-loop DBS control schemes [29].

We developed a computational model of the cortical-basal ganglia-thalamus circuit in the 6-OHDA-lesioned rat model of PD, including a closed-loop connection from the thalamus to cortex [58]. The properties of the model were validated extensively, including responses evoked by CTX stimulation in Str, STN, GPe, and GPi model neurons with experimental PSTHs [55]. The model accounted for key differences observed in the response patterns between the normal and PD states, and the firing rates and patterns observed in the normal and PD states were consistent with those in experimental studies in rats. We used the model to quantify the frequency-dependent effects of STN DBS on low-frequency oscillatory activity in model neurons. The model accounted for the frequency-dependent effects of STN DBS with high frequencies being more effective at suppressing the pathological oscillatory activity compared to low frequencies. This rodent version of the BG network was coupled with a novel genetic algorithm to optimize temporal patterns of STN DBS [21].

5.6 Opportunities to Improve Network Models of the BG for Optimization of DBS

5.6.1 Symptom-Specific Network Models

The major motor symptoms of PD are rest tremor, rigidity, bradykinesia, postural instability, and gait imbalance. Other non-motor symptoms include cognitive dysfunctions, speech deficits, mood disorders, etc. The major motor symptoms are not expressed to an equal extent across individuals with PD, i.e., some persons have more rest tremor compared to bradykinesia and vice versa [80]. Therefore, there is a need to develop symptom-specific network models of the BG circuit rather than generic ones. For example, synchronous oscillations of the BG neurons in the beta band (13–35 Hz) are correlated with bradykinesia, whereas theta-band oscillations (3–7 Hz) might be related to tremor symptoms [19]. STN and GPi are preferred DBS targets during expression of all PD motor symptoms, whereas ventral intermediate nucleus of the TH is the preferred target in tremor-dominant PD [5]. Thus, computational models of the BG network exhibiting symptom-specific neural signatures can be used to optimize DBS therapy for individual patients.

5.6.2 Dynamic-Specific Network Models

The symptoms of PD are dynamic with current state such as movements vs. rest, as well as dependent on medication status and time since last dose. Therefore, improved network models of BG should capture these dynamics in PD symptoms with respect to a trigger signal indicating an event such as initiation of movement. Desynchronization of beta-band oscillations is a common phenomenon observed during the onset of movements [61].

5.6.3 Improved Cost Function for Model-Based Optimization

The challenges with current DBS technology are side effects and high energy expenditure due to the necessity of HF stimulation. Model-based optimization might enable identification of DBS parameters that reduce PD symptoms with reduced energy consumption [16]. Activation of internal capsule fibers is implicated as the major cause for DBS-induced side effects. Hence, there is a need to develop model-based proxies indicating induced side effects for a given set of DBS parameters. Further, parameters might also be optimized to reduce stimulation-evoked side effects. For example, reducing the average frequency of stimulation tends to reduce side effects [83]. Additionally, current model-based optimization approaches do not restore the rates and patterns of neural activity seen during healthy conditions (Fig. 5.2). Therefore, future model-based optimization studies can optimize DBS patterns such that neural activity during PD+HF STN DBS condition is same as the healthy state. We envision future DBS technology to be neurorestorative, i.e., restore neural activity seen in the healthy BG, and models that faithfully represent neural activity and the response to DBS will be a critical element of developing such therapies.

Acknowledgments This work was supported by a grant from the US National Institutes of Health (NIH R37 NS040894).

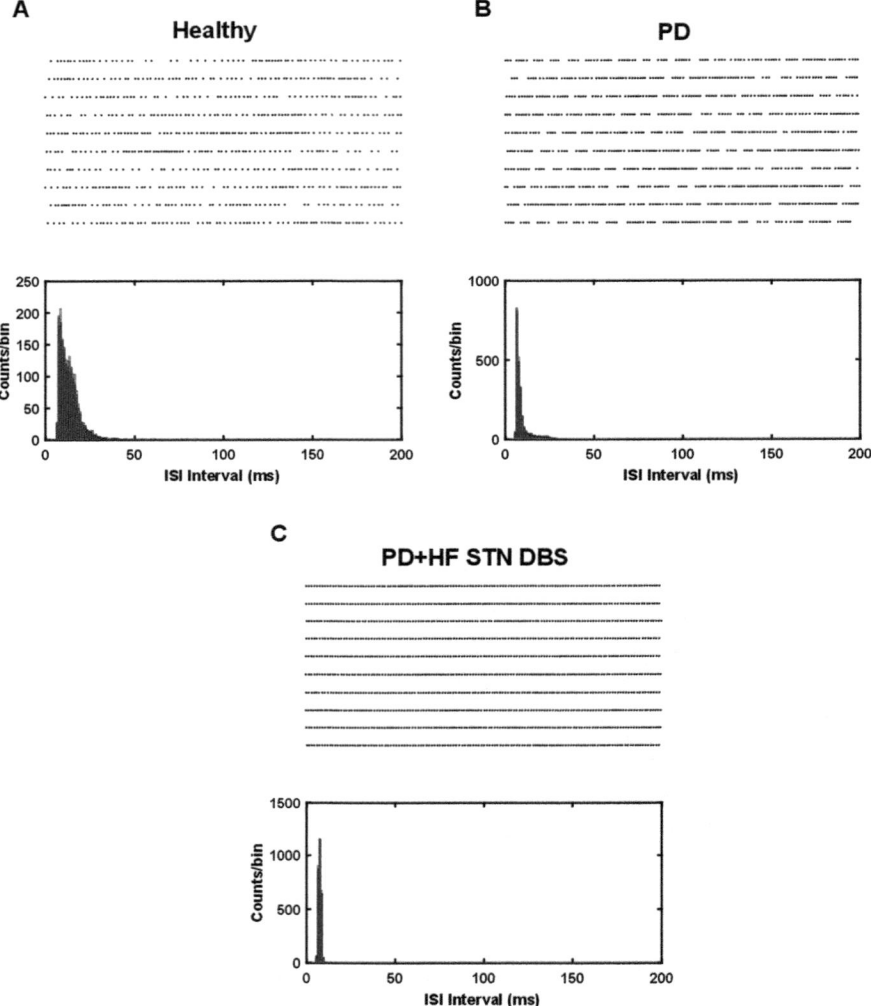

Fig. 5.2 Model-based neural response (rasterogram and interspike interval histogram of GPi neurons) during three different conditions (**a**) healthy, (**b**) Parkinson's disease (PD), and (**c**) PD+high-frequency subthalamic nucleus deep brain stimulation (HF STN DBS). Note the asynchronous firing in healthy condition, while PD state exhibits synchronous burst firing. PD+HF STN DBS results in entrainment of model neurons at the frequency of STN DBS. Note the neural response during PD+HF STN DBS state is not the same as the healthy condition. In other words, HF STN DBS does not restore the response seen during normal condition instead induces a different neural response which results in suppression of PD motor symptoms. Therefore, future DBS technology can be neurorestorative where the goal is to optimize stimulation patterns such that neural response during PD+HF STN DBS is similar to the healthy condition

References

1. Adamchic I, Hauptmann C, Barnikol UB, Pawelczyk N, Popovych O, Barnikol TT, Silchenko A, Volkmann J, Deuschl G, Meissner WG (2014) Coordinated reset neuromodulation for Parkinson's disease: proof-of-concept study. Mov Disord 29:1679–1684

2. Agarwal R, Sarma SV (2012) The effects of DBS patterns on basal ganglia activity and thalamic relay. J Comput Neurosci 33:151–167

3. Agid Y (1987) Biochemistry of neurotransmitters in Parkinson's disease. Mov Disord 2:166–230

4. Albin RL, Young AB, Penney JB (1989) The functional anatomy of basal ganglia disorders. Trends Neurosci 12:366–375

5. Anderson D, Beecher G, Ba F (2017) Deep brain stimulation in Parkinson's disease: new and emerging targets for refractory motor and nonmotor symptoms. Parkinson's Dis 2017:5124328

6. Anderson ME, Postupna N, Ruffo M (2003) Effects of high-frequency stimulation in the internal globus pallidus on the activity of thalamic neurons in the awake monkey. J Neurophysiol 89:1150–1160

7. Arefin MS (2012). Performance analysis of single-site and multiple-site deep brain stimulation in basal ganglia for Parkinson's disease. In: Electrical & Computer Engineering (ICECE), 2012 7th international conference on IEEE, p 149–152

8. Ashkan K, Rogers P, Bergman H, Ughratdar I (2017) Insights into the mechanisms of deep brain stimulation. Nat Rev Neurol 13:548–554

9. Baudrexel S, Witte T, Seifried C, von Wegner F, Beissner F, Klein JC, Steinmetz H, Deichmann R, Roeper J, Hilker R (2011) Resting state fMRI reveals increased subthalamic nucleus–motor cortex connectivity in Parkinson's disease. NeuroImage 55:1728–1738

10. Benabid A-L, Pollak P, Louveau A, Henry S, De Rougemont J (1987) Combined (thalamotomy and stimulation) stereotactic surgery of the VIM thalamic nucleus for bilateral Parkinson disease. Stereotact Funct Neurosurg 50:344–346

11. Bergman H, Wichmann T, Karmon B, DeLong M (1994) The primate subthalamic nucleus. II. Neuronal activity in the MPTP model of parkinsonism. J Neurophysiol 72:507–520

12. Beurrier C, Congar P, Bioulac B, Hammond C (1999) Subthalamic nucleus neurons switch from single-spike activity to burst-firing mode. J Neurosci 19:599–609

13. Bezard E, Boraud T, Bioulac B, Gross CE (1999) Involvement of the subthalamic nucleus in glutamatergic compensatory mechanisms. Eur J Neurosci 11:2167–2170

14. Bin-Mahfoodh M, Hamani C, Sime E, Lozano AM (2003) Longevity of batteries in internal pulse generators used for deep brain stimulation. Stereotact Funct Neurosurg 80:56–60

15. Bolam J, Hanley J, Booth P, Bevan M (2000) Synaptic organisation of the basal ganglia. J Anat 196:527–542

16. Brocker DT, Swan BD, So RQ, Turner DA, Gross RE, Grill WM (2017) Optimized temporal pattern of brain stimulation designed by computational evolution. Sci Transl Med 9:eaah3532

17. Brocker DT, Swan BD, Turner DA, Gross RE, Tatter SB, Koop MM, Bronte-Stewart H, Grill WM (2013) Improved efficacy of temporally non-regular deep brain stimulation in Parkinson's disease. Exp Neurol 239:60–67

18. Bronte-Stewart H, Barberini C, Koop MM, Hill BC, Henderson JM, Wingeier B (2009) The STN beta-band profile in Parkinson's disease is stationary and shows prolonged attenuation after deep brain stimulation. Exp Neurol 215:20–28

19. Brown P (2003) Oscillatory nature of human basal ganglia activity: relationship to the pathophysiology of Parkinson's disease. Mov Disord 18:357–363

20. Calabresi P, Picconi B, Tozzi A, Ghiglieri V, Di Filippo M (2014) Direct and indirect pathways of basal ganglia: a critical reappraisal. Nat Neurosci 17:1022–1030

21. Cassar IR, Titus ND, Grill WM (2017) An improved genetic algorithm for designing optimal temporal patterns of neural stimulation. J Neural Eng 14:066013

22. Cassidy M, Mazzone P, Oliviero A, Insola A, Tonali P, Lazzaro VD, Brown P (2002) Movement-related changes in synchronization in the human basal ganglia. Brain 125:1235–1246
23. Chen Y, Wang J, Wei X, Deng B, Che Y (2011) Particle swarm optimization of periodic deep brain stimulation waveforms. In: Control Conference (CCC), 2011 30th Chinese IEEE, p 754–757
24. Chu H-Y, McIver EL, Kovaleski RF, Atherton JF, Bevan MD (2017) Loss of hyperdirect pathway cortico-subthalamic inputs following degeneration of midbrain dopamine neurons. Neuron 95:1306–1318. e1305
25. Cui G, Jun SB, Jin X, Pham MD, Vogel SS, Lovinger DM, Costa RM (2013) Concurrent activation of striatal direct and indirect pathways during action initiation. Nature 494:238–242
26. Daneshzand M, Faezipour M, Barkana BD (2017) Computational stimulation of the basal ganglia neurons with cost effective delayed Gaussian waveforms. Front Comput Neurosci 11:73
27. Deuschl G, Schade-Brittinger C, Krack P, Volkmann J, Schäfer H, Bötzel K, Daniels C, Deutschländer A, Dillmann U, Eisner W (2006) A randomized trial of deep-brain stimulation for Parkinson's disease. N Engl J Med 355:896–908
28. Dorval AD, Kuncel AM, Birdno MJ, Turner DA, Grill WM (2010) Deep brain stimulation alleviates parkinsonian bradykinesia by regularizing pallidal activity. J Neurophysiol 104:911–921
29. Dunn EM, Lowery MM (2013) Simulation of PID control schemes for closed-loop deep brain stimulation. In: Neural Engineering (NER), 2013 6th international IEEE/EMBS conference on IEEE, p 1182–1185
30. Ebert M, Hauptmann C, Tass PA (2014) Coordinated reset stimulation in a large-scale model of the STN-GPe circuit. Front Comput Neurosci 8:154
31. Fan D, Wang Q (2015) Improving desynchronization of parkinsonian neuronal network via triplet-structure coordinated reset stimulation. J Theor Biol 370:157–170
32. Feng X-J, Shea-Brown E, Greenwald B, Kosut R, Rabitz H (2007) Optimal deep brain stimulation of the subthalamic nucleus—a computational study. J Comput Neurosci 23:265–282
33. Filion M (1991) Abnormal spontaneous activity of globus pallidus neurons in monkeys with MPTP-induced parkinsonism. Brain Res 547:140–144
34. Gerfen CR, Wilson CJ (1996) Chapter II: The basal ganglia. In: Swanson LW, Björklund A, Hokfelt T (eds) Handbook of chemical neuroanatomy, Vol. 12: Integrated systems of the CNS, Part III. Elsevier Science Publishers, New York, pp 371–468
35. Glynn G, Ahmad S (2002) Three-dimensional electrophysiological topography of the rat corticostriatal system. J Comp Physiol A Neuroethol Sens Neural Behav Physiol 188:695–703
36. Gorzelic P, Schiff S, Sinha A (2013) Model-based rational feedback controller design for closed-loop deep brain stimulation of Parkinson's disease. J Neural Eng 10:026016
37. Grill WM, Cantrell MB, Robertson MS (2008) Antidromic propagation of action potentials in branched axons: implications for the mechanisms of action of deep brain stimulation. J Comput Neurosci 24:81–93
38. Group D-BSfPsDS (2001) Deep-brain stimulation of the subthalamic nucleus or the pars interna of the globus pallidus in Parkinson's disease. N Engl J Med 2001:956–963
39. Guiot G, Brion S(1953) Traitement des mouvements anormaux par la coagulation pallidale-Technique et resultats. In: Revue Neurologique MASSON EDITEUR 120 BLVD SAINT-GERMAIN, 75280 PARIS 06, FRANCE, p 578–580
40. Guo Y, Rubin JE (2011) Multi-site stimulation of subthalamic nucleus diminishes thalamocortical relay errors in a biophysical network model. Neural Netw 24:602–616
41. Guo Y, Rubin JE, McIntyre CC, Vitek JL, Terman D (2008) Thalamocortical relay fidelity varies across subthalamic nucleus deep brain stimulation protocols in a data-driven computational model. J Neurophysiol 99:1477–1492
42. Hahn PJ, McIntyre CC (2010) Modeling shifts in the rate and pattern of subthalamopallidal network activity during deep brain stimulation. J Comput Neurosci 28:425–441

43. Hashimoto T, Elder CM, Okun MS, Patrick SK, Vitek JL (2003) Stimulation of the subthalamic nucleus changes the firing pattern of pallidal neurons. J Neurosci 23:1916–1923

44. Hassani O-K, Mouroux M, Feger J (1996) Increased subthalamic neuronal activity after nigral dopaminergic lesion independent of disinhibition via the globus pallidus. Neuroscience 72:105–115

45. Hassler R, Riechert T (1954) Indikationen und Lokalisationsmethode der gezielten Hirnoperationen. Nervenarzt 25:441–447

46. Hauptmann C, Tass PA (2010) Restoration of segregated, physiological neuronal connectivity by desynchronizing stimulation. J Neural Eng 7:056008

47. Heimer G, Bar-Gad I, Goldberg JA, Bergman H (2002) Dopamine replacement therapy reverses abnormal synchronization of pallidal neurons in the 1-methyl-4-phenyl-1, 2, 3, 6-tetrahydropyridine primate model of parkinsonism. J Neurosci 22:7850–7855

48. Herrington TM, Cheng JJ, Eskandar EN (2016) Mechanisms of deep brain stimulation. J Neurophysiol 115:19–38

49. Hollerman JR, Grace AA (1992) Subthalamic nucleus cell firing in the 6-OHDA-treated rat: basal activity and response to haloperidol. Brain Res 590:291–299

50. Holt AB, Netoff TI (2014) Origins and suppression of oscillations in a computational model of Parkinson's disease. J Comput Neurosci 37:505–521

51. Holt AB, Wilson D, Shinn M, Moehlis J, Netoff TI (2016) Phasic burst stimulation: a closed-loop approach to tuning deep brain stimulation parameters for Parkinson's disease. PLoS Comput Biol 12:e1005011

52. Humphries MD, Stewart RD, Gurney KN (2006) A physiologically plausible model of action selection and oscillatory activity in the basal ganglia. J Neurosci 26:12921–12942

53. Jankovic J, Rajput AH, McDermott MP, Perl DP (2000) The evolution of diagnosis in early Parkinson disease. Arch Neurol 57:369–372

54. Kang G, Lowery MM (2013) Interaction of oscillations, and their suppression via deep brain stimulation, in a model of the cortico-basal ganglia network. IEEE Trans Neural Syst Rehabil Eng 21:244–253

55. Kita H, Kita T (2011) Cortical stimulation evokes abnormal responses in the dopamine-depleted rat basal ganglia. J Neurosci 31:10311–10322

56. Krack P, Batir A, Van Blercom N, Chabardes S, Fraix V, Ardouin C, Koudsie A, Limousin PD, Benazzouz A, LeBas JF (2003) Five-year follow-up of bilateral stimulation of the subthalamic nucleus in advanced Parkinson's disease. N Engl J Med 349:1925–1934

57. Kühn AA, Kempf F, Brücke C, Doyle LG, Martinez-Torres I, Pogosyan A, Trottenberg T, Kupsch A, Schneider G-H, Hariz MI (2008) High-frequency stimulation of the subthalamic nucleus suppresses oscillatory β activity in patients with Parkinson's disease in parallel with improvement in motor performance. J Neurosci 28:6165–6173

58. Kumaravelu K, Brocker DT, Grill WM (2016) A biophysical model of the cortex-basal ganglia-thalamus network in the 6-OHDA lesioned rat model of Parkinson's disease. J Comput Neurosci 40:207–229

59. Kuncel AM, Grill WM (2004) Selection of stimulus parameters for deep brain stimulation. Clin Neurophysiol 115:2431–2441

60. Laitinen LV, Bergenheim AT, Hariz MI (1992) Leksell's posteroventral pallidotomy in the treatment of Parkinson's disease. J Neurosurg 76:53–61

61. Levy R, Ashby P, Hutchison WD, Lang AE, Lozano AM, Dostrovsky JO (2002) Dependence of subthalamic nucleus oscillations on movement and dopamine in Parkinson's disease. Brain 125:1196–1209

62. Limousin P, Krack P, Pollak P, Benazzouz A, Ardouin C, Hoffmann D, Benabid A-L (1998) Electrical stimulation of the subthalamic nucleus in advanced Parkinson's disease. N Engl J Med 339:1105–1111

63. Liu C, Wang J, Deng B, Wei X, Yu H, Li H, Fietkiewicz C, Loparo KA (2016) Closed-loop control of tremor-predominant parkinsonian state based on parameter estimation. IEEE Trans Neural Syst Rehabil Eng 24:1109–1121

64. Liu C, Wang J, Li H, Lu M, Deng B, Yu H, Wei X, Fietkiewicz C, Loparo KA (2017) Closed-loop modulation of the pathological disorders of the basal ganglia network. IEEE Trans Neural Netw Learn Syst 28:371–382
65. Lourens MA, Schwab BC, Nirody JA, Meijer HG, van Gils SA (2015) Exploiting pallidal plasticity for stimulation in Parkinson's disease. J Neural Eng 12:026005
66. Lozano AM, Lang AE, Galvez-Jimenez N, Miyasaki J, Duff J, Hutchison W, Dostrovsky JO (1995) Effect of GPi pallidotomy on motor function in Parkinson's disease. Lancet 346:1383–1387
67. Lysyansky B, Popovych OV, Tass PA (2013) Optimal number of stimulation contacts for coordinated reset neuromodulation. Front Neuroengineering 6:5
68. Magill P, Bolam J, Bevan M (2001) Dopamine regulates the impact of the cerebral cortex on the subthalamic nucleus–globus pallidus network. Neuroscience 106:313–330
69. Mallet N, Pogosyan A, Márton LF, Bolam JP, Brown P, Magill PJ (2008) Parkinsonian beta oscillations in the external globus pallidus and their relationship with subthalamic nucleus activity. J Neurosci 28:14245–14258
70. McIntyre CC, Grill WM, Sherman DL, Thakor NV (2004) Cellular effects of deep brain stimulation: model-based analysis of activation and inhibition. J Neurophysiol 91:1457–1469
71. Meyers R (1942) Surgical interruption of the pallidofugal fibers. Its effect on the syndrome of paralysis agitans and technical considerations in its application. NY State J Med 42:317–325
72. Moran RJ, Mallet N, Litvak V, Dolan RJ, Magill PJ, Friston KJ, Brown P (2011) Alterations in brain connectivity underlying beta oscillations in parkinsonism. PLoS Comput Biol 7:e1002124
73. Moro E, Esselink R, Xie J, Hommel M, Benabid A, Pollak P (2002) The impact on Parkinson's disease of electrical parameter settings in STN stimulation. Neurology 59:706–713
74. Nambu A, Tokuno H, Hamada I, Kita H, Imanishi M, Akazawa T, Ikeuchi Y, Hasegawa N (2000) Excitatory cortical inputs to pallidal neurons via the subthalamic nucleus in the monkey. J Neurophysiol 84:289–300
75. Nambu A, Tokuno H, Takada M (2002) Functional significance of the cortico–subthalamo–pallidal 'hyperdirect' pathway. Neurosci Res 43:111–117
76. Pan HS, Walters JR (1988) Unilateral lesion of the nigrostriatal pathway decreases the firing rate and alters the firing pattern of globus pallidus neurons in the rat. Synapse 2:650–656
77. Pirini M, Rocchi L, Sensi M, Chiari L (2009) A computational modelling approach to investigate different targets in deep brain stimulation for Parkinson's disease. J Comput Neurosci 26:91
78. Plenz D, Kital ST (1999) A basal ganglia pacemaker formed by the subthalamic nucleus and external globus pallidus. Nature 400:677
79. Popovych OV, Lysyansky B, Rosenblum M, Pikovsky A, Tass PA (2017) Pulsatile desynchronizing delayed feedback for closed-loop deep brain stimulation. PLoS One 12:e0173363
80. Rajput A, Sitte H, Rajput A, Fenton M, Pifl C, Hornykiewicz O (2008) Globus pallidus dopamine and Parkinson motor subtypes clinical and brain biochemical correlation. Neurology 70:1403–1410
81. Ray N, Jenkinson N, Wang S, Holland P, Brittain J, Joint C, Stein J, Aziz T (2008) Local field potential beta activity in the subthalamic nucleus of patients with Parkinson's disease is associated with improvements in bradykinesia after dopamine and deep brain stimulation. Exp Neurol 213:108–113
82. Rizzone M, Lanotte M, Bergamasco B, Tavella A, Torre E, Faccani G, Melcarne A, Lopiano L (2001) Deep brain stimulation of the subthalamic nucleus in Parkinson's disease: effects of variation in stimulation parameters. J Neurol Neurosurg Psychiatry 71:215–219
83. Rosa M, Arlotti M, Ardolino G, Cogiamanian F, Marceglia S, Di Fonzo A, Cortese F, Rampini PM, Priori A (2015) Adaptive deep brain stimulation in a freely moving parkinsonian patient. Mov Disord 30:1003–1005
84. Rubin JE, Terman D (2004) High frequency stimulation of the subthalamic nucleus eliminates pathological thalamic rhythmicity in a computational model. J Comput Neurosci 16:211–235

85. Smith Y, Beyan M, Shink E, Bolam J (1998) Microcircuitry of the direct and indirect pathways of the basal ganglia. Neurosci-Oxford 86:353–388
86. So RQ, Kent AR, Grill WM (2012) Relative contributions of local cell and passing fiber activation and silencing to changes in thalamic fidelity during deep brain stimulation and lesioning: a computational modeling study. J Comput Neurosci 32:499–519
87. Soares J, Kliem MA, Betarbet R, Greenamyre JT, Yamamoto B, Wichmann T (2004) Role of external pallidal segment in primate parkinsonism: comparison of the effects of 1-methyl-4-phenyl-1, 2, 3, 6-tetrahydropyridine-induced parkinsonism and lesions of the external pallidal segment. J Neurosci 24:6417–6426
88. Su F, Wang J, Deng B, Li H (2015a) Effects of deep brain stimulation amplitude on the basal-ganglia-thalamo-cortical network. In: Control and Decision Conference (CCDC), 2015 27th Chinese IEEE, p 4049–4053
89. Su F, Wang J, Deng B, Wei X-L, Chen Y-Y, Liu C, Li H-Y (2015b) Adaptive control of Parkinson's state based on a nonlinear computational model with unknown parameters. Int J Neural Syst 25:1450030
90. Summerson SR, Aazhang B, Kemere C (2015) Investigating irregularly patterned deep brain stimulation signal design using biophysical models. Front Comput Neurosci 9:78
91. Tass PA, Qin L, Hauptmann C, Dovero S, Bezard E, Boraud T, Meissner WG (2012) Coordinated reset has sustained aftereffects in parkinsonian monkeys. Ann Neurol 72:816–820
92. Terman D, Rubin JE, Yew A, Wilson C (2002) Activity patterns in a model for the subthalamopallidal network of the basal ganglia. J Neurosci 22:2963–2976
93. Vila M, Perier C, Feger J, Yelnik J, Faucheux B, Ruberg M, Raisman-Vozari R, Agid Y, Hirsch E (2000) Evolution of changes in neuronal activity in the subthalamic nucleus of rats with unilateral lesion of the substantia nigra assessed by metabolic and electrophysiological measurements. Eur J Neurosci 12:337–344
94. Wang J, Nebeck S, Muralidharan A, Johnson MD, Vitek JL, Baker KB (2016) Coordinated reset deep brain stimulation of subthalamic nucleus produces long-lasting, dose-dependent motor improvements in the 1-methyl-4-phenyl-1, 2, 3, 6-tetrahydropyridine non-human primate model of parkinsonism. Brain Stimul 9:609–617
95. Wang R, Wang J, Chen Y, Deng B, Wei X (2011) A new deep brain stimulation waveform based on PWM. In: Biomedical Engineering and Informatics (BMEI), 2011 4th international conference on IEEE, p 1815–1819
96. Weaver FM, Follett K, Stern M, Hur K, Harris C, Marks WJ, Rothlind J, Sagher O, Reda D, Moy CS (2009) Bilateral deep brain stimulation vs best medical therapy for patients with advanced Parkinson disease: a randomized controlled trial. JAMA 301:63–73
97. Wichmann T, Bergman H, Starr PA, Subramanian T, Watts RL, DeLong MR (1999) Comparison of MPTP-induced changes in spontaneous neuronal discharge in the internal pallidal segment and in the substantia nigra pars reticulata in primates. Exp Brain Res 125:397–409
98. Wichmann T, Soares J (2006) Neuronal firing before and after burst discharges in the monkey basal ganglia is predictably patterned in the normal state and altered in parkinsonism. J Neurophysiol 95:2120–2133

Chapter 6
Neural Synchronization in Parkinson's Disease on Different Time Scales

Sungwoo Ahn, Choongseok Park, and Leonid L. Rubchinsky

Abstract Parkinson's disease is marked by an elevated neural synchrony in the cortico-basal ganglia circuits in the beta frequency band. This elevated synchrony has been associated with Parkinsonian hypokinetic symptoms. The application of recently developed synchronization analysis techniques allows us to investigate the temporal dynamics of synchrony on different time scales. The results of this analysis are summarized here, revealing highly variable dynamics of synchronized neural activity on multiple time scales and its association with disease.

Keywords Parkinson's disease · Neural oscillations · Neural synchronization · Desynchronization · Intermittency · Beta-band oscillations

6.1 Beta-Band Oscillations and Synchronization in Parkinson's Disease

Synchronized rhythms of neural activity are widely observed phenomena in the brain and have been studied quite extensively because of their correlations with multiple functions and dysfunctions of neural systems. Neural synchronization plays a crucial role in perception, cognition, and memory, among other processes

S. Ahn (✉)
Department of Mathematics, East Carolina University, Greenville, NC, USA
e-mail: ahns15@ecu.edu

C. Park
Department of Mathematics, North Carolina A&T State University, Greensboro, NC, USA

L. L. Rubchinsky
Department of Mathematical Sciences, Indiana University-Purdue University Indianapolis, Indianapolis, IN, USA

Stark Neurosciences Research Institute, Indiana University School of Medicine, Indianapolis, IN, USA
e-mail: lrubchin@iupui.edu

© Springer Nature Switzerland AG 2019
V. Cutsuridis (ed.), *Multiscale Models of Brain Disorders*, Springer Series in Cognitive and Neural Systems 13, https://doi.org/10.1007/978-3-030-18830-6_6

(reviewed in [8, 11, 12]). Abnormalities of neural synchrony (such as excessively strong or excessively weak synchrony) have been related to the symptoms of several neurological and psychiatric disorders (reviewed in [21, 29, 31]).

In particular, abnormalities of neural oscillations and synchrony have been observed in the cortico-basal ganglia circuits in Parkinson's disease. Parkinson's disease is a major neurodegenerative disorder characterized by chronic dopamine deficiency resulting in a set of movement-related as well as other symptoms (see, e.g., [20] and references therein). The loss of dopamine in Parkinson's disease directly affects the basal ganglia, a group of subcortical nuclei which are, among other things, involved in the neural control of movement. The landmark of Parkinson's disease is overall slowness of movement. This hypokinetic behavior involves bradykinesia and akinesia (slowness of ongoing movement/inability to start new movement) and rigidity (stiffness of joints). Another frequent symptom is rest tremor whose biological mechanisms are probably different from those of hypokinesia.

Parkinsonian pathophysiology is marked by increased oscillatory and synchronous activity in the beta frequency band in cortical and basal ganglia circuits. Over the past two decades, many studies have reported on the relationship between excessive oscillations and synchronization in the beta-band and hypokinetic motor deficits in humans with Parkinson's disease and in animal models of this disorder (reviewed in, e.g., [10, 13, 28, 30]).

Even though Parkinsonian brain expresses elevated beta-band synchrony, this synchrony is still relatively mild [22, 30]. It changes in time, and most conventional methods of synchronization estimation miss a complex picture of temporal dynamics of synchrony. However, several techniques for the analysis of the dynamics of synchrony reveal different temporal patterns of synchrony on different time scales (see below). Here we review recent progress in the development of these synchronization analysis techniques and their applications to Parkinsonian neurodynamics.

6.2   Synchronization on Different Time Scales

There are many definitions of synchronization, but the common theme is coordination of the temporal aspects of the oscillations, usually because of the coupling between underlying oscillations. From the observational standpoint, synchronization is inherently non-instantaneous phenomenon, and this is what distinguishes it from a random and non-repetitive coincidence of some oscillatory features of two signals [24]. This leads to the difficulty in estimation of synchrony over short time scales. To make this discussion more specific, let us focus here on phase synchronization.

Phase domain is an appropriate way to analyze weakly synchronized neural signals [14, 17, 18, 24, 32]. As the coupling strength increases from low to moderate values, synchrony may be observed in the phase domain, while the amplitudes of oscillations remain uncorrelated. The phase may provide a more sensitive metric to explore moderately synchronized neural activity. The phase can be extracted from oscillatory data in different ways including the use of the Hilbert transformation. Let us assume that two phases are extracted from two signals: φ_1 and φ_2. Then one can compute a fairly standard phase-locking (or phase synchrony) index γ:

$$\gamma = \left\| \frac{1}{N} \sum_{j=1}^{N} e^{i\,\theta_j} \right\|^2 ,$$

where $\theta_j = \varphi_1(t_j) - \varphi_2(t_j)$ is the phase difference, t_j are the times of data points, and N is the number of data points. This phase-locking index varies from 0 (no phase synchrony) to 1 (perfect phase synchrony). This phase-locking index was used to study neural synchronization of widely varying strength, but it naturally provides an average strength of phase synchrony.

However, behavior and synchrony, which helps to mediate it, usually vary in time, so there is a question of how synchronization varies in time. To address this problem, one may estimate a phase-locking index over time window of certain fixed length. But for confident evaluation of synchrony, one needs to observe it for a relatively long time. One can approach this issue statistically [14], by constructing surrogates to evaluate phase-locking significance. Depending on the time scale used in the analysis, there will be different temporal synchrony patterns [14]. This is not an artifact of the analysis. Depending on which time scale is physiological, synchrony may be significant or not, not only statistically but physiologically.

Decreasing the length of the analysis time window necessarily degrades statistical power. The window size must be long enough for powerful statistics and yet short enough for high temporal resolution. Importantly, this may render short analysis windows impractical. However, if there is an overall synchrony, one can consider how the system gets to a synchronized state and leaves it in time (synchronized state needs to be appropriately defined). This approach was recently developed in [1, 27] and can describe the differences in the temporal structure of synchronization and desynchronization events for the systems with similar overall level of phase-locking. This is important given that the average neural synchrony is frequently not very strong. The underlying network of presumably weakly coupled oscillators spends a substantial fraction of time in the desynchronized state, which justifies the focus on desynchronization episodes.

We will briefly describe one possible realization of this approach by using the first-return map analysis to quantify deviations from the synchronized state, provided that the data exhibit some synchrony on the average. Whenever the phase of one signal crosses zero level from negative to positive values, we record the phase of the other signal, generating a set of consecutive values $\{\phi_i\}$, $i = 1, \ldots, N$. These ϕ_i represent the phase difference between two signals. After determining the most frequent value of ϕ_i, all the phases are shifted accordingly (for different episodes under consideration) so that averaging across different episodes (with potentially different phase shifts) is possible. Thus, this approach is not concerned with the

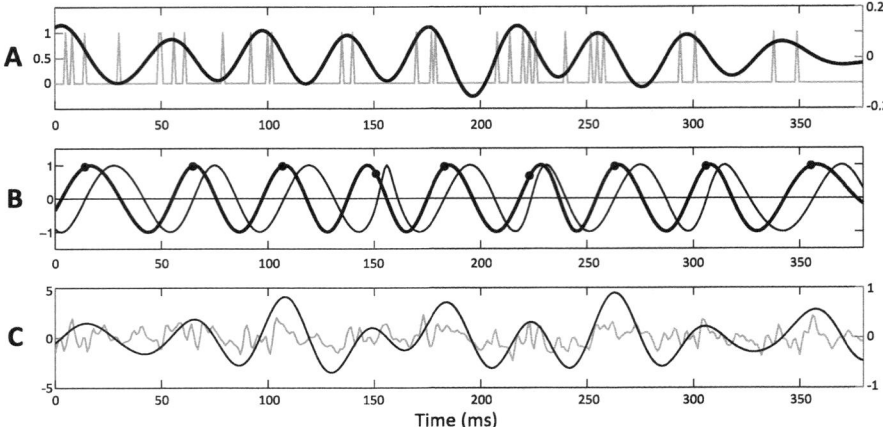

Fig. 6.1 An example of a synchronized episode. (**a**) Raw (thin line) and band-pass (10–30 Hz) filtered spiking signal (thick line). (**c**) Raw local field potential (LFP, grey line) and band-pass filtered signal (black line). (**b**) The sines of the phases of the filtered spiking (thick curve) and the filtered LFP (thin curve) signals. The amplitude information is lost here, but the phase information is preserved. Dots indicate the phases of the filtered spiking signal whenever the phase of filtered LFP signal crosses 0 upward. (Adapted from [22])

value of the phase shift between signals, but rather with the maintenance of the constant phase shift (phase-locking) (see Fig. 6.1).

Dynamics is considered as desynchronized if the phase difference deviates from the preferred phase difference by more than certain amount ($\pi/2$ was used in several studies). The duration of the desynchronized episodes is measured in cycles of the oscillations. Thus, if the phase difference deviates from the preferred phase difference by more than $\pi/2$ once, then the duration of the desynchronized episode is one. If it deviates twice, then the duration is two, etc. This approach distinguishes between many short desynchronizations, few long desynchronizations and the possibilities in between even if they all yield the same average synchrony strength.

We will describe the results of application of these techniques to the studies of Parkinson's disease neurophysiology in the next section.

6.3   Synchronization in Parkinson's Disease on Different Time Scales

The application of the synchronization variations analysis to the subcortical intraoperative recordings from Parkinsonian patients indicates that the phase-locking index γ exhibits substantial variation in time. Figure 6.2 illustrates that the question of

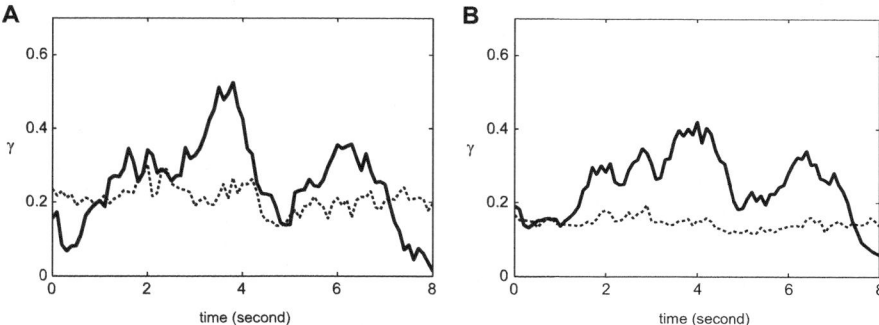

Fig. 6.2 Temporal dynamics of synchronous activity depends on the analysis window length. Black curve is the phase-locking index γ computed over a short time window with duration of 1 s (**a**) and 1.5 s (**b**). Dotted curve is the 95% significance level estimate, obtained from surrogate data. (Adapted from [28])

whether the dynamics is synchronous or not depends on the time scale. The data used here are the spiking activity and LFP in the subthalamic nucleus of patients with advanced Parkinson's disease (subthalamic nucleus LFP is likely to reflect pallidal input to the subthalamic nucleus, so this synchrony may be indicative of pallidal-subthalamic relationship or input-output relationship for subthalamic nucleus, discussed in [22]). The values of γ depend on the analysis window length. This is natural, for long time windows one expects to see less time variability, while for shorter time window, one has a better temporal resolution and more time variability as well as less powerful statistics.

The synchronization between motor cortices in Parkinson's disease follows a similar pattern [7]. The time course of synchrony (as evaluated in this time-dependent manner) in cortical and basal ganglia networks happens to be correlated in a manner specific to pairs of EEG electrodes over motor and prefrontal cortical areas, pointing to potentially global functional interaction between cortex and the basal ganglia in Parkinson's disease, when elevated synchrony in one network may impact synchronous dynamics in another one.

The synchronization index γ considered above does not inform about the fine temporal structure of synchrony because the analysis window length is not very small. A window size of 1 s corresponds to ~20 cycles of beta oscillations. Exploration of synchrony patterns on finer time scales is possible with techniques described in the previous section. This approach revealed the intermittent nature of activity in Parkinsonian brain and specifically the fine temporal structure of beta oscillations: synchronous states are interrupted by frequent, but short desynchronizations (see Fig. 6.3). The signals go out of phase for just one cycle of oscillations more often than for two or a larger number of cycles in the basal ganglia [22, 25].

Beta-band activity in Parkinson's disease is associated with hypokinetic symptoms. Another prominent Parkinsonian symptom is rest tremor. It is confined to other frequency band (3–8 Hz), is expressed independently of beta activity [26], and

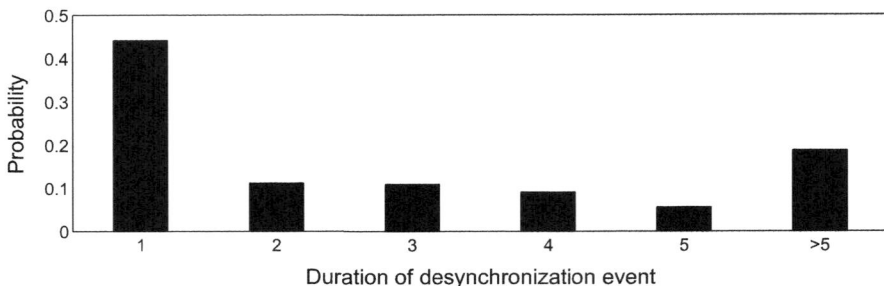

Fig. 6.3 The histogram of desynchronization event durations (measured in cycles of oscillations). For the duration >5, all durations longer than 5 are pooled together. The data are for the window length for the computation of synchronization index γ equal to 1.5 s. (Adapted from [22])

is likely to have separate network mechanism. However, it also expresses temporal patterning of neural (or neuro-muscular) synchrony, which is different on different time scales [15, 16]. High temporal variability of all these pathological neural synchronized dynamics may be related to the fact that these oscillations per se may be normal, but being overexpressed synchrony leads to pathological symptoms (see discussion in [22, 23, 28]).

6.4 Modeling Patterns of Neural Synchrony

Complex interactions within and between nuclei may be responsible for intermittently synchronized beta rhythms in Parkinson's disease. Experiments suggest that two nuclei, subthalamic nucleus and external globus pallidus, may form a key substrate for the synchronous rhythms in the Parkinsonian basal ganglia. Different models of subthalamo-pallidal circuits of basal ganglia were used to study how properties of neurons interact with network properties to generate synchronized rhythms (e.g., [19]). A potential problem with this approach is that getting moderate synchrony in coupled oscillators is easy, so matching frequency and average synchrony strength may not be constraining enough.

Matching the temporal patterning of synchrony, especially on the very short time scales, may be an effective tool to match the modeling and experimental data. While there may be many different characteristics of dynamics to match between model and experiment, for the phenomena where synchrony is important, matching synchronous patterns allows to match the phase space of model and real systems. Since basal ganglia synchronous dynamics is very intermittent, matching synchrony patterns in the model and experiment ensures some similarity between large areas of the phase space of the model and real systems. Thus, the mechanisms of synchronized oscillatory activity considered in the model may be able to produce the experimentally observed dynamics (see discussion in [9, 22]).

We have used the matching of synchrony patterns as a tool to find parameter values for the models of the cortical and basal ganglia networks, which would generate realistic synchronous beta-band oscillations. This approach suggested that Parkinsonian state of basal ganglia networks at rest is on the border of synchronized and non-synchronized activity [23]. This new approach showed how Parkinsonian synchronized beta oscillations may be promoted by the simultaneous action of both cortical (or some other) and subthalamo-pallidal network mechanisms [6]. It also showed that some proposed types of deep brain stimulation in Parkinson's disease may be potentially either effective [25] or ineffective [9] in pathological synchrony suppression. The latter is an interesting observation because it emphasizes that effectiveness of suppression of pathological synchrony may depend on how this synchrony is patterned in time.

6.5 Conclusions

Synchronization is inherently non-instantaneous phenomenon, and its temporal dynamics depend on the time scale used for the analysis. Even the question of whether there is a statistically significant synchrony or not depends on the time scale under consideration [14]. Many neural synchrony phenomena and behaviors that they mediate are short-lived and non-stationary. Thus, the temporal aspects of neural synchrony are likely to be important.

In particular, this is the case for the pathological neural synchrony in Parkinson's disease. As we described here, the basal ganglia express specific temporal patterns of the synchronous beta-band activity [22, 28], which are likely to be dopamine-dependent [23]. Parkinsonian tremor expresses different synchrony patterns on different temporal scales [15]. Temporal variations of the beta-band synchrony in Parkinson's disease are also observed in cortico-basal ganglia interactions, and temporal variability of synchronous patterns in cortical and basal ganglia circuits is related [6, 7]. And we also would like to note that not only temporal but spatial aspects are relevant to Parkinsonian physiology too [33].

The temporal patterning of synchrony phenomena are not confined to Parkinson's disease. Alterations of synchrony patterns on short time scales have been observed in addicted brain [4] and even in the coordination of brainstem-regulated respiratory rhythm and cardiac rhythm in with disease vs. healthy states [5].

Frequently observed patterning of neural synchrony on the very short time scales (the interruption of synchronous dynamics by potentially numerous but predominantly very short desynchronizations) may be a generic property of neural circuits in the brain even in the healthy state [2, 4]. It may be grounded in the very basic properties of the excitability of neural membranes [3]. However, the quantitative differences between patterning of neural synchrony may be related to relatively mild but behaviorally significant changes in the underlying network.

Acknowledgments This paper was supported by the ICTSI/Indiana University Health–IU School of Medicine Strategic Research Initiative, the National Science Foundation Grant DMS 1813819, and National Science Foundation Grant HRD 1700199.

References

1. Ahn S, Park C, Rubchinsky LL (2011) Detecting the temporal structure of intermittent phase locking. Phys Rev E 84(1–2):016201
2. Ahn S, Rubchinsky LL (2013) Short desynchronization episodes prevail in synchronous dynamics of human brain rhythms. Chaos 23(1):013138
3. Ahn S, Rubchinsky LL (2017) Potential mechanisms and functions of intermittent neural synchronization. Front Comput Neurosci 11:44
4. Ahn S, Rubchinsky LL, Lapish CC (2014a) Dynamical reorganization of synchronous activity patterns in prefrontal cortex – hippocampus networks during behavioral sensitization. Cereb Cortex 24(10):2553–2561
5. Ahn S, Solfest J, Rubchinsky LL (2014b) Fine temporal structure of cardiorespiratory synchronization. Am J Physiol Heart Circ Physiol 306(5):H755–H763
6. Ahn S, Zauber SE, Worth RM, Rubchinsky LL (2016) Synchronized beta-band oscillations in a model of the globus pallidus – subthalamic nucleus network under external input. Front Comput Neurosci 10:134
7. Ahn S, Zauber SE, Worth RM, Witt T, Rubchinsky LL (2015) Interaction of synchronized dynamics in cortical and basal ganglia networks in Parkinson's disease. Eur J Neurosci 42(5):2164–2171
8. Buzsáki G, Schomburg EW (2015) What does gamma coherence tell us about inter-regional neural communication? Nat Neurosci 18:484–489
9. Dovzhenok A, Park C, Worth RM, Rubchinsky LL (2013) Failure of delayed feedback deep brain stimulation for intermittent pathological synchronization in Parkinson's disease. PLoS One 8(3):e58264
10. Eusebio A, Brown P (2009) Synchronisation in the beta frequency-band – the bad boy of parkinsonism or an innocent bystander? Exp Neurol 217(1):1–3
11. Fell J, Axmacher N (2011) The role of phase synchronization in memory processes. Nat Rev Neurosci 12(2):105–118
12. Fries P (2015) Rhythms for cognition: communication through coherence. Neuron 88(1):220–235
13. Hammond C, Bergmann H, Brown P (2007) Pathological synchronization in Parkinson's disease: networks, models and treatments. Trends Neurosci 30(7):357–364
14. Hurtado JM, Rubchinsky LL, Sigvardt KA (2004) Statistical method for detection of phase-locking episodes in neural oscillations. J Neurophysiol 91(4):1883–1898
15. Hurtado JM, Rubchinsky LL, Sigvardt KA, Wheelock VL, Pappas CTE (2005) Temporal evolution of oscillations and synchrony in GPi/muscle pairs in Parkinson's disease. J Neurophysiol 93(3):1569–1584
16. Hurtado JM, Rubchinsky LL, Sigvardt KA (2006) The dynamics of tremor networks in Parkinson's disease. In: Bezard E (ed) Recent breakthroughs in basal ganglia research. Nova Science publishers, New York, pp 249–266
17. Lachaux JP, Rodriguez E, Martinerie J, Varela FJ (1999) Measuring phase synchrony in brain signals. Hum Brain Mapp 8(4):194–208
18. Le Van Quyen M, Bragin A (2007) Analysis of dynamic brain oscillations: methodological advances. Trends Neurosci 30(7):365–373
19. Mandali A, Rengaswamy M, Chakravarthy VS, Moustafa AA (2015) A spiking basal ganglia model of synchrony, exploration and decision making. Front Neurosci 9:191

20. Obeso JA, Stamelou M, Goetz CG, Poewe W, Lang AE, Weintraub D, Burn D, Halliday GM, Bezard E, Przedborski S, Lehericy S, Brooks DJ, Rothwell JC, Hallett M, DeLong MR, Marras C, Tanner CM, Ross GW, Langston JW, Klein C, Bonifati V, Jankovic J, Lozano AM, Deuschl G, Bergman H, Tolosa E, Rodriguez-Violante M, Fahn S, Postuma RB, Berg D, Marek K, Standaert DG, Surmeier DJ, Olanow CW, Kordower JH, Calabresi P, Schapira AHV, Stoessl AJ (2017) Past, present, and future of Parkinson's disease: a special essay on the 200th anniversary of the shaking palsy. Mov Disord 32:1264–1310
21. Oswal A, Brown P, Litvak V (2013) Synchronized neural oscillations and the pathophysiology of Parkinson's disease. Curr Opin Neurol 26(6):662–670
22. Park C, Worth RM, Rubchinsky LL (2010) Fine temporal structure of beta oscillations synchronization in subthalamic nucleus in Parkinson's disease. J Neurophysiol 103(5):2707–2716
23. Park C, Worth RM, Rubchinsky LL (2011) Neural dynamics in parkinsonian brain: the boundary between synchronized and nonsynchronized dynamics. Phys Rev E 83:042901
24. Pikovsky A, Rosenblum M, Kurths J (2001) Synchronization: a universal concept in nonlinear sciences. Cambridge University Press, Cambridge
25. Ratnadurai-Giridharan S, Zauber SE, Worth RM, Witt T, Ahn S, Rubchinsky LL (2016) Temporal patterning of neural synchrony in the basal ganglia in Parkinson's disease. Clin Neurophysiol 127(2):1743–1745
26. Ray NJ, Jenkinson N, Wang S, Holland P, Brittain JS, Joint C, Stein JF, Aziz T (2008) Local field potential beta activity in the subthalamic nucleus of patients with Parkinson's disease is associated with improvements in bradykinesia after dopamine and deep brain stimulation. Exp Neurol 213(1):108–113
27. Rubchinsky LL, Ahn S, Park C (2014) Dynamics of desynchronized episodes in intermittent synchronization. Front Phys 2:38
28. Rubchinsky LL, Park C, Worth RM (2012) Intermittent neural synchronization in Parkinson's disease. Nonlinear Dyn 68(3):329–346
29. Schnitzler A, Gross J (2005) Normal and pathological oscillatory communication in the brain. Nat Rev Neurosci 6(4):285–296
30. Stein E, Bar-Gad I (2013) Beta oscillations in the cortico-basal ganglia loop during parkinsonism. Exp Neurol 245:52–59
31. Uhlhaas PJ, Singer W (2010) Abnormal neural oscillations and synchrony in schizophrenia. Nat Rev Neurosci 11(2):100–113
32. Varela F, Lachaux JP, Rodriguez E, Martinerie J (2001) The brainweb: phase synchronization and large-scale integration. Nat Rev Neurosci 2(4):229–239
33. Zaidel A, Spivak A, Grieb B, Bergman H, Israel Z (2010) Subthalamic span of β oscillations predicts deep brain stimulation efficacy for patients with Parkinson's disease. Brain 133:2007–2021

Chapter 7
Obsessive-Compulsive Tendencies and Action Sequence Complexity: An Information Theory Analysis

Mustafa Zeki, Fuat Balcı, Tutku Öztel, and Ahmed A. Moustafa

Abstract Obsessive-compulsive disorder (OCD) is a psychiatric condition that is primarily associated with anxiety provoking repetitive thoughts (i.e., obsessions) and actions that are manifested to neutralize the resultant anxiety (i.e., compulsions). Interestingly, OCD patients continue compulsive behaviors (e.g., repeatedly rechecking if the door is locked) although they are typically aware of the irrationality of these behaviors. This suggests that compulsive behaviors have habit-like features. We predicted that the motor actions (e.g., sequence of goalless key presses) would deviate from randomness in individuals with stronger obsessive-compulsive (OC) tendencies and thus expected to observe more rigid sequential action patterns in these individuals (e.g., pressing keys according to a motif). We applied entropy theory approach, defined as the rate of change of information in a given sequence, to test this hypothesis. We collected two different types of sequential behavioral data from healthy individuals and scored their obsessive-compulsive tendencies based on the Padua Inventory. In the first method, we asked participants to press one of the two buttons sequentially. In the second method, participants were asked to mark one of the four different options sequentially (on a multiple-choice optic form). The behavioral characterization was carried out by quantifying the entropy

M. Zeki (✉)
Department of Mathematics, College of Engineering and Technology,
American University of the Middle East, Egaila, Kuwait
e-mail: mustafa.zeki@aum.edu.kw

F. Balcı
Research Center for Translational Medicine & Department of Psychology, Koç University,
Istanbul, Turkey
e-mail: Fbalci@ku.edu.tr

T. Öztel
Research Center for Translational Medicine, Koç University, Istanbul, Turkey

A. A. Moustafa
School of Social Sciences and Psychology & Marcs Institute for Brain and Behaviour, Western
Sydney University, Sydney, New South Wales, Australia
e-mail: a.moustafa@westernsydney.edu.au

© Springer Nature Switzerland AG 2019
V. Cutsuridis (ed.), *Multiscale Models of Brain Disorders*, Springer Series in
Cognitive and Neural Systems 13, https://doi.org/10.1007/978-3-030-18830-6_7

67

in the sequence of two sets of behavioral data using the Shannon metric entropy and Lempel-Ziv complexity measures. Our results revealed a negative relationship between the degree of washing tendencies and the level of information contained in action sequences. These results held only for the data collected with key presses and not for the choice sequences in the paper-pencil task. Based on these results, we conclude that the behavioral rigidity observed in the form of compulsive actions may generalize to some other behaviors of the individual.

Keywords Shannon metric entropy · Lempel-Ziv complexity ·
Obsessive-compulsive disorder · Entropy · Information theory · Action
sequences

7.1 Introduction

Obsessive-compulsive disorder (OCD) is a mental disorder characterized by two main elements: obsessions and compulsions. Obsessions are unwanted, disruptive, and often repetitive thoughts, feelings, and sensations. Compulsions are repetitive motor (checking) or mental (counting) acts aimed at seizing the observed obsessions or anxiety that they cause. Attempts to inhibit the compulsions forcibly cause steep increases in the current level of anxiety often resulting in the patient to repeatedly engage in compulsions. Furthermore, some compulsions (such as a specific way of washing hands) carry the features of fixed action patterns as rigid sequential behavioral programs (e.g., Berridge et al. [2]). In this work, we investigated if such rigid features of behavioral sequences are generalized to other actions that are not part of an individual's repertoire of compulsions.

There are several tools that originate from the information theory that are suitable to capture the degree of rigidity in this type of action sequences. The complexity of a symbolic sequence is defined as the rate of change of information in that sequence. Intuitively, it is the measure of the information content of a given sequence. The Shannon metric entropy (ME) uses the probabilities of symbolic "words" of a given length in the calculation of the entropy [21]. Consequently, ME is sensitive to both the frequency and ordering of the symbols in the sequence. The Lempel-Ziv complexity (LZC) measures the information content by constituting a dictionary of distinct "words" in increasing lengths by applying an algorithm that chooses the shortest "word" that previously did not appear or that is not in the dictionary yet. It is the resultant number of distinct words in the dictionary that is used to calculate the LZC of the given symbolic sequence. In data-scientific terms, LZC measures the compressibility of a given sequence. However, both of these entropy measures are defined for the sequences of infinite length. Their application for very short symbolic sequences has been the subject of many recent studies [1, 8, 12]. In one of these studies, Lesne et al. [11] compared many types of entropy measures in terms of their goodness of approximation and found out that Shannon ME and LZC [10] were superior to other measures for both sequences with low and high entropy values.

In the current study, we applied these information theoretic approaches to characterize the relationship between the obsessive-compulsive tendencies of healthy individuals and the degree of rigidity in their action sequences without goals. In one of the tasks, participants were asked to repeatedly choose between four options on an optic form, whereas in the other task, they were asked to repeatedly press one of the two buttons on a game pad. The obsessive-compulsive tendencies of participants were quantified based on the Padua Inventory (washing, rumination, impulses, checking, precision subscales). The information content (complexity) of the sequential actions was quantified based on tools that are most suitable for short-length symbolic data [11] (i.e., Shannon ME and LZC). Our results revealed significant relationships between the complexity of behavioral sequences and primarily the washing tendencies.

7.2 Methods

7.2.1 Participants

Participants were recruited from Koç University undergraduate students (mean age 20.2) in exchange for course credit. Sample was composed of 69 participants (39 in the first group, 30 in the second group). The data of three participants in the first group (instructed to "respond as they wished") were excluded from the analysis since two of them did not fully complete the Padua Inventory and one of them could not complete the game pad task because of a technical problem. There were 35 females and 60 right-handed individuals in the remaining sample that was composed of 66 participants. We applied the median absolute deviation (MAD) method introduced in Leys et al. [13] on the ME and LZC scores to exclude those participants who had executed the experiment with excessive ME and LZC scores due to reasons such as pressing the same button repeatedly for long periods during testing. In particular, an individual was regarded as an outlier if the corresponding data of the individual is more than three median absolute deviation away from the median. Based on this exclusion criterion, ten participants were excluded from further analysis. Consequently, 56 participants were included in the analysis (26 females, 52 right-handed, mean age 20.12). All the procedures were approved by the Koç University Institutional Review Board.

7.2.2 Material

Stimuli were presented on an iMac computer using Psychophysics Toolbox extensions [4, 7, 16] that were run on Matlab R2015b. Key presses were collected with a Logitech F710 Wireless game pad that was connected to the iMac.

7.2.3   Procedure

Participants first completed the paper-pencil (multiple-choice optic form) and game pad choice tasks (order counterbalanced). Following the completion of these tasks, participants filled out the Padua Inventory.

7.2.4   Game Pad Choice Task

Participants were asked to press one of the two keys on a game pad for 1000 times as they wished (in the first group of participants) or randomly (in the second group of participants). Contingent upon each key press, a square appeared on the screen and remained until the participant released the pressed key. With the presentation of the square, a beep sound was played during the key press for up to 500 ms. Participants could press the keys at any rate. The comparison of the entropy scores between the two groups of participants did not reveal a significant difference either for ME ($t(54) = 0.24$, $p = 0.81$) or LZC ($t(54) = 0.71$, $p = 0.48$). Based on the lack of significant differences, data were pooled between the two groups.

7.2.5   Paper-Pencil (Multiple Choice Form) Task

Participants were given an empty multiple-choice optic form that contained items with four choice options per item. They were then asked to choose one of the options (A, B, C, and D) for 150 times starting with the first item. Each sheet was composed of three columns, each of which contained 25 items. Participants filled out two sheets and were instructed to place each sheet in their sight while filling out the other one. Each sheet had print only on one side. Participants were instructed that there was no right or wrong answer of their choices, thus they could answer as they wished (in the first group of participants) or randomly (in the second group of participants). The comparison of the entropy scores between the two groups did not reveal a significant difference either for ME ($t(54) = -0.62$, $p = 0.54$) or LZC ($t(54) = -1.13$, $p = 0.26$). Based on the lack of significant differences, data were pooled between the two groups of participants.

7.2.6   Scales

Following behavioral testing, participants filled out the Padua Inventory ([19, 23]; see Beşiroğlu et al. [3] for Turkish). The Padua Inventory measures the severity of OCD symptoms and contains five subscales: checking, washing, rumination, impulses, and precision [23].

7.2.7 Data Analysis

Information Theory Analysis: *Entropy Calculations for Symbolic Data*
   Shannon metric entropy (ME): For a symbolic sequence of infinite length, Shannon ME ([21], also known as *box entropy*) of order n is defined as

$$H_n = - \sum_{S_n} p\,(S_n)\,\log_2 p\,(S_n)\,.$$

The sum is taken among all substrings of length n, S_n with probability of $p(S_n)$. This quantity measures the amount of information contained in words of length n.

Intuitively, it is the amount of information or the amount of uncertainty in a given sequence. The following example (modified from Claude Shannon [21]) gives a better intuition for understanding the formula above for $n = 1$. Let us consider a machine producing one of the letters A, B, C, and D at each step. The entropy is now defined as the least number of questions to be asked on the average to guess the next letter. If all letters appear in the same frequency, their probabilities will be equal to $p(S_1)$, where S_1 is one of the four letters. In order to guess the next letter, it is sufficient to ask two questions for each letter on the average. Namely, one can ask, "Is it A or B?". If the answer is "Yes," one can then ask "Is it A?" to determine if it is A or B. Otherwise, one would ask "Is it C?" to determine if it is C or D. For each choice, the number of questions to be asked is two. Hence the weighted average would be

$$0.25 \times 2 + 0.25 \times 2 + 0.25 \times 2 + 0.25 \times 2 = 2.$$

If the letters do not appear with equal probabilities, the least number of questions to be asked is reduced. If the new probabilities are 0.5 for A, 0.25 for B, and 0.125 for C and D, one asks questions with respect to the probabilities of the letters. Since A appears in majority of the cases, finding it first reduces the average number of questions to be asked to determine the next outcome. One first asks "Is it A?" if the answer is "Yes," the task is done; If not, one then asks "Is it B?" since B appears most frequently after A. If the answer is "Yes," the task is done. If not, one now asks "Is it C?". Now the expected number of questions to be asked is

$$0.50 \times 1 + 0.25 \times 2 + 0.125 \times 3 + 0.125 \times 3 = 1.750.$$

Interestingly, Shannon found out that the number of questions to be asked can be determined using the probabilities of the symbols. The number of questions to be asked to determine a given letter appears to be $\log_2(1/p(S_1))$ with S_1 is one of

A, B, C, or D. Indeed, in the above example, the formula yields $\log_2(1/0.5) = 1$, $\log_2(1/0.25) = 2$, $\log_2(1/0.125) = 3$, $\log_2(1/0.125) = 3$, respectively.

The entropy rate h_r, is given by the limit

$$h_r = \lim_{n \to \infty} \frac{H_n}{n}$$

which can be approximated by the difference

$$h_{sap} := H_{n+1} - H_n$$

for some $n > 0$.

Lempel-Ziv complexity (LZC): The LZC [10] also calculates the information content of a given data sequence. It is usually used to compress large data sets by creating a dictionary of the recurrent pieces in their first appearances. Instead of recording redundant words, their index from the dictionary is recorded. In finding the LZC of a given symbolic sequence of length N and k symbols, the sequence is broken into substrings in such a way that the next word would be the shortest substring that has not appeared before. For example, given the binary sequence

$$10110011001010$$

the sequence of the substrings would be

$$1 * 0 * 11 * 00 * 110 * 01 * 010.$$

Now, the number of words in the partition is used to calculate the LZC of the data. If we denote the number of substrings with N_s, the LZC is given by

$$L = \lim_{N \to \infty} L_N \, with \, L_N = \frac{N_s \left(1 + \log_k N_s\right)}{N}$$

which can be approximated with $h_{lap} = L_N$ for some $N > 0$. The relation of LZC with Shannon ME is given as follows [11]:

$$\lim_{N \to \infty} L_N = \frac{h_r}{\ln 2 \ln k} \, or \, h_r = L \ln 2 \ln k$$

Shannon ME is the measure of mean uncertainty of predicting a substring of length n. Then, the entropy rate would be the change in the uncertainty. Intuitively, this is translated as a change in the information added by increasing "word" length by 1. On the other hand, the LZC of a symbolic sequence of length N is a measure of distinct patterns in a given sequence. In their elegant work, Lesne et al. [11] considered various entropy approximations for very short symbolic sequences with correlated (deterministic) and uncorrelated (random) dynamics. Among the ones considered, they found that the above approximations for Shannon and LZC gave the best results in terms of the difference with the entropy value of the untruncated sequences. In particular, the LZC approximation h_{lap} gave much better results for sequences with low entropy dynamics; and Shannon ME approximation h_{sap} was equally good for sequences with low and high entropy values.

7.3 Results

Tables 7.1 and 7.2 summarize the results of the correlation and multiple linear regression analyses, respectively. The entropy scores (both LZC and ME) were negatively correlated with the washing tendencies. These results held for ME even when the analyses were constrained to the first 150 choices, namely, to the number of choices in the paper-pencil task. In addition, both ME and LZC of the first 150 items of the data were inversely correlated with overall Padua total score. The ME and LZC of the first 150 choices also showed inverse correlation with the impulses subscale score. We would like to stress that in two independent pilot studies we had previously conducted, the relationship between the entropy scores and the degree of washing tendency corroborated our current findings, increasing our confidence primarily in our results in relation to the washing tendency.

No significant correlations were found for the data gathered with the paper-pencil multiple-choice task.

The correlation results also indicate that OCD subscale scores are strongly correlated among each other ($p < 0.001$ and $r > 0.65$ for all subscales). In addition, individual entropy approximations by both ME and LZ algorithms were strongly correlated for both the paper-pencil multiple-choice ($p < 0.001$, $r = 0.83$) and game pad data ($p < 0.001$, $r = 0.76$).

We next performed a stepwise multiple linear regression analyses with each Padua subscale as the predictor variable and (a) LZC, (b) ME, (c) LZC (first 150 items), and (d) ME (first 150 items) as the predicted variables. Criterion to add terms into the regression was restricted to P-values that were less than 0.05. Each predicted variable was entered separately in the regression analysis. The final models are shown in Table 7.2. The LZC and ME of the entire data were predicted

Table 7.1 Correlations between Padua subscales and entropy measures for the game pad data. Only statistically significant results are shown (Spearman's rho and P-value)

Participants ($n = 56$)	Lempel-Ziv compl. (LZC) scores	Metric entropy (ME) scores	Lempel-Ziv compl. (LZC) (first 150 items)	(ME) (first 150 items)
Padua total score	–	–	−0.34*	−0.27*
Washing subscale	−0.32*	−0.32*	–	−0.27*
Impulses subscale	–	–	−0.51***	−0.34*
Precision subscale	–	–	–	–

*indicates $p < 0.05$, ** indicates $p < 0.01$, *** indicates $p < 0.001$

Table 7.2 Stepwise multiple linear regression predicting entropy scores using each Padua subscale score as a predictor; only statistically significant results are shown

Participants ($n = 56$)	LZC[a]		ME[b]		LZC (first 150 items)[c]		ME (first 150 items)[d]	
	b	t	b	t	b	t	b	t
Washing subscale	-0.32*	-2.48	-0.3*	-2.3	-	-	-0.34**	-2.7
Impulses subscale	-	-	-	-	-0.5***	-4.31	-0.49***	-3.3
Checking subscale	-	-	-	-	-	-	0.31*	2.05

[a] Adjusted $R^2 = 0.085$; $F_{(1,54)} = 6.14$; $P = 0.01$
[b] Adjusted $R^2 = 0.074$; $F_{(1,54)} = 5.37$; $P = 0.024$
[c] Adjusted $R^2 = 0.243$; $F_{(1,54)} = 18.6$; $P < 0.001$
[d] Adjusted $R^2 = 0.196$; $F_{(3,52)} = 5.47$; $P < 0.002$
indicates $p < 0.05$, ** indicates $p < 0.01$, *** indicates $p < 0.001$; standardized slopes are presented

by washing scores. The predictions that were constrained to the initial segment of the behavioral sequences (equivalent to the number of choices in the paper-pencil task) also spanned the impulses (for both ME and LZC) and washing tendencies (for ME); increase in both features predicted higher rigidity during this segment of the data. Interestingly, however, checking tendencies predicted lower rigidity during the initial segment of the data with ME. We would like to note that these positive results should be interpreted with caution as they resulted from exploratory analyses.

7.4 Discussion

This study tested if obsessive-compulsive features in a subclinical group of partic-ipants predicted lower action syntax complexity (higher action sequence rigidity) measured by the entropy in the sequence of key presses/choices. Our results pointed at a consistent relationship between the washing tendencies and action sequence complexity in the game pad data (that required key presses). The paper-pencil data did not result in any significant relationship with any of the variables. To our knowledge, this is the first time investigation of the entropy of the behavioral data in relation to a psychological construct(s).

The fact that the participants had immediate access to their previous choices in the paper-pencil task is a possible reason for the lack of the relationship between entropy of the choices gathered from this task and the Padua scores. That is, having access to the previous choices may have altered the entropy of the obtained data due to the induction of a common behavioral tendency in participants with low and high obsessive-compulsive tendencies. Emergence of common behavioral patterning might have masked the inherent differences between participants with differential obsessive-compulsive tendencies. Another reason for the absence of the correlation between the entropy of choices and the corresponding Padua scores may be the size of the data collected in the paper-pencil task (150 items each). In order to address this possibility, we calculated the correlations of total Padua and Padua subscale scores with the entropy estimated from the initial segment (150 items) of the game pad data. Though not as strong as the entire game pad data, the entropy scores of the initial segment of the game pad data still resulted in strong correlations with the Padua scores (total Padua, impulses, and washing scores; see Table 7.1). Furthermore, our stepwise regression analysis showed the predictive value of washing tendencies for action sequence rigidity that also included/replaced by impulses tendencies when the data were restricted to the initial segment of testing. These results suggest that game pad task might be more sensitive to obsessive-compulsive tendencies primarily due to the task characteristics (key press vs. paper-pencil, two vs. four choice options) rather than task-independent peripheral factors such as the sample size. Further studies are needed to empirically address these different possibilities.

These consistent relationships with specifically washing tendencies might be due to different neural mechanisms that underlie washing compared to other forms of the obsessive-compulsive tendencies (e.g., checking and rumination). Neuroimaging studies suggest that the provocation of washing, checking, and hoarding symptoms in OCD patients activate different brain areas to varying degrees in addition to some overlap (e.g., for review see Taylor et al. [22]). For instance, Mataix-Cols et al. [15] asked participants to imagine different scenarios that related either to washing, checking, or hoarding symptoms. The provocation of washing symptom in OCD patients resulted in greater activity of the caudate nucleus and several cortical regions (e.g., ventromedial prefrontal cortex) than healthy controls, whereas higher activation of other brain areas were observed with the provocation of checking and hoarding symptoms. In the same study, different Padua subscale scores (i.e., washing, checking, hoarding) were found to be correlated with the activity of different brain regions. Supporting the heterogeneity of OCD [14], different subtypes of OCD also respond differentially to different treatments (for review see Taylor et al. [22]) and exhibit differential decision-making deficits [9]. A higher degree of overlap between the neural circuits that are related to the performance in our task and washing tendencies compared to other symptoms of OCD might underlie the specific relations observed in this study.

The evolutionary and neuroethological approaches to OCD suggest that compulsions might just be exaggerated forms of otherwise adaptive responses such as washing for hygiene/prevention of disease and checking for safety, some of which might be (at least partially) genetically coded [5, 6, 17]. To this end, Rapoport [17] emphasized possible links between washing and grooming, essentially conceptualizing washing as a fixed action pattern. Similar exaggerated or ritualized forms of natural behaviors were also observed in other animals, and they can be treated with medications effective in human OCD (e.g., Rapoport et al. [18], Seksel and Lindeman [20]). These approaches to OCD also support the specific relations we have observed in our study combined with previous findings gathered from rodent studies.

Overall, our results suggest that particularly washing tendencies but also the total scores and impulses tendencies obtained in a widely used OCD inventory can predict the action sequence syntax complexity even in a subclinical sample. We expect that the extension of this study to clinical populations would reveal more robust findings. Finally, our findings emphasize the importance of using information theory analysis to understand the behavioral features of the psychiatric conditions.

References

1. Adami C, Cerf NJ (2000) Physical complexity of symbolic sequences. Phys D: Nonlinear Phenom 137(1):62–69
2. Berridge KC, Aldridge JW, Houchard KR, Zhuang X (2005) Sequential super-stereotypy of an instinctive fixed action pattern in hyper-dopaminergic mutant mice: a model of obsessive-compulsive disorder and Tourette's. BMC Biol 3(2):4
3. Beşiroğlu L, Ağargün MY, Boysan M, Eryonucu B, Güleç M, Selvi Y (2005) The assessment of obsessive-compulsive symptoms: the reliability and validity of the Padua inventory in a Turkish population. Turk Psikiyatri Derg 16:179–189

4. Brainard DH (1997) The psychophysics toolbox. Spat Vis 10:433–436
5. Brune M (2006) The evolutionary psychology of obsessive-compulsive disorder: the role of cognitive metarepresentation. Perspect Biol Med 49(3):317–329
6. Insel TR (1988) Obsessive-compulsive disorder: a neuroethological perspective. Psychopharmacol Bull 24(3):365–369
7. Kleiner M, Brainard D, Pelli D, Ingling A, Murray R, Broussard C (2007) What's new in psychtoolbox-3. Perception 36(14):1–16
8. Labate D, La Foresta F, Morabito G, Palamara I, Morabito FC (2013) Entropic measures of EEG complexity in Alzheimer's disease through a multivariate multiscale approach. IEEE Sensors J 13(9):3284–3292
9. Lawrence NS, Wooderson S, Mataix-Cols D, David R, Speckens A, Phillips ML (2006) Decision making and set shifting impairments are associated with distinct symptom dimensions in obsessive-compulsive disorder. Neuropsychology 20(4):409–419
10. Lempel A, Ziv J (1976) On the complexity of finite sequences. IEEE Trans Inf Theory IT-22(1):75–81
11. Lesne A, Blanc J, Pezard L (2009) Entropy estimation of very short symbolic sequences. Phys Rev E 79(4):046208
12. Lesne A (2014) Shannon entropy: a rigorous notion at the crossroads between probability, information theory, dynamical systems and statistical physics. Math Struct Comput Sci 24:3
13. Leys C, Ley C, Klein O, Bernard P, Licata L (2013) Detecting outliers: do not use standard deviation around the mean, use absolute deviation around the median. J Exp Soc Psychol 49(4):764–766
14. Lochner C, Stein DJ (2003) Heterogeneity of obsessive-compulsive disorder: a literature review. Harv Rev Psychiatry 11(1):13–132
15. Mataix-Cols D, Wooderson S, Lawrence N, Brammer MJ, Speckens A, Phillips ML (2004) Distinct neural correlates of washing, checking, and hoarding symptom dimensions in obsessive-compulsive disorder. Arch Gen Psychiatry 61(6):564–576
16. Pelli DG (1997) The VideoToolbox software for visual psychophysics: transforming numbers into movies. Spat Vis 10:437–442
17. Rapoport JL (1989) The biology of obsessions and compulsions. Sci Am 260:63–69
18. Rapoport JL, Ryland DH, Kriete M (1992) Drug treatment of canine acral lick. An animal model of obsessive-compulsive disorder. Arch Gen Psychiatry 49(7):517–521
19. Sanavio E (1988) Obsessions and compulsions: the Padua inventory. Behav Res Ther 26(2):169–177
20. Seksel K, Lindeman MJ (2001) Use of clomipramine in treatment of obsessive-compulsive disorder, separation anxiety and noise phobia in dogs: a preliminary, clinical study. Aust Vet J 79(4):252–256
21. Shannon CE (1948) A mathematical theory of communication. Bell Syst Tech J 27:379–423
22. Taylor S, McKay D, Abramowitz JS (2008) Making sense of obsessive-compulsive disorder. In: Abramowitz JS, McKay D, Taylor S (eds) Clinical handbook of obsessive-compulsive disorder and related problems. The Johns Hopkins University Press, Baltimore
23. Van Oppen P, Hoekstra RJ, Emmelkamp PM (1995) The structure of obsessive-compulsive symptoms. Behav Res Ther 33(1):15–23

Part II
Cognitive Disorders

Chapter 8
Cortical Disinhibition, Attractor Dynamics, and Belief Updating in Schizophrenia

Rick A. Adams

Abstract Genetic and pharmacological evidence implicates N-methyl-D-aspartate receptor (NMDAR) dysfunction in the pathophysiology of schizophrenia. Dysfunction of this key receptor – if localised to inhibitory interneurons – could cause a net disinhibition of cortex and increase in 'noise'. These effects can be computationally modelled in a variety of ways: by reducing the precision in Bayesian models of behaviour, by estimating neuronal excitability changes in schizophrenia from evoked responses, or – as described in detail here – by modelling abnormal belief updating in a probabilistic inference task. Features of belief updating in schizophrenia include greater updating to unexpected evidence, lower updating to consistent evidence, and greater stochasticity in responding. All of these features can be explained by a loss of stability of 'attractor states' in cortex and the representations they encode. Indeed, a hierarchical Bayesian model of belief updating indicates that subjects with schizophrenia have a consistently increased 'belief instability' parameter. This instability could be a direct result of cortical disinhibition: this hypothesis should be explored in future studies.

Keywords Schizophrenia · Psychosis · Computational · Beads task · Excitation-inhibition balance · Bayesian

A key challenge in schizophrenia spectrum research is understanding how deficiencies in synaptic function in general and N-methyl-D-aspartate receptor (NMDAR) functioning in particular in the disorder – implied by genetic studies [1] and pharmacological models of psychosis that use the NMDAR antagonist ketamine [2] – impact on neural dynamics at the circuit, network, and whole-brain levels. A further challenge is understanding how these changes in neural dynamics then affect brain computations and behaviour.

R. A. Adams (✉)
Institute of Cognitive Neuroscience, University College London, London, UK

Division of Psychiatry, University College London, London, UK
e-mail: rick.adams@ucl.ac.uk

© Springer Nature Switzerland AG 2019
V. Cutsuridis (ed.), *Multiscale Models of Brain Disorders*, Springer Series in
Cognitive and Neural Systems 13, https://doi.org/10.1007/978-3-030-18830-6_8

NMDARs are located on both excitatory pyramidal cells and inhibitory interneurons, although different receptor subtypes are distributed differentially on different populations [3]. NMDAR dysfunction could therefore impact inhibitory interneurons or pyramidal cells to differing degrees: in 'subjects with a diagnosis of schizophrenia' (Scz), there is evidence that the former are more strongly affected [4], resulting in a net loss of inhibitory (relative to excitatory) transmission. This disinhibited state is also known as 'increased E/I balance'.

The consequences of increased E/I balance in Scz can be modelled in a variety of ways. One approach is to assume that this disinhibited state causes a decrease in precision (increase in variance) of the states that neural circuits encode: especially circuits at higher levels of the hierarchy, where there is most evidence for inhibitory dysfunction. One can then model these effects as the loss of precision of prior beliefs within a hierarchical Bayesian model of behaviour, as prior beliefs are most affected by loss of precision at the top of a hierarchical model. This approach has shown that numerous perceptual or behavioural phenomena in Scz can be modelled in this way, e.g. dysfunction of smooth pursuit eye movements, resistance to visual illusions, etc. [5].

Another approach is to ignore behaviour altogether and just model neural responses. One of the best-validated electroencephalographic (EEG) findings in Scz is a reduction in the mismatch negativity [6]. The mismatch negativity is the difference in averaged EEG deflection in response to an oddball stimulus (e.g. a high tone following a series of low tones) compared to that following the standard. In Scz, there is less of a difference between EEG responses to oddballs and standards than there is in controls, and E/I balance could contribute to this.

Dynamic causal modelling (DCM) estimates how the activity in neural populations (e.g. pyramidal cells or interneurons) in connected brain areas evolves in response to some input (e.g. a sensory stimulus) according to the parameters of the system (e.g. the degree of disinhibition of pyramidal cells within areas or the strength of connections between areas). DCM of mismatch negativity responses of Scz and their first-degree relatives and healthy controls indicate that both Scz and their relatives have (i) an increase in disinhibition in the (right inferior) prefrontal area involved in the mismatch response and (ii) a reversal of the usual increase in excitability in response to oddballs (seen in controls) in that source [7]. This not only supports the notion of cortical disinhibition in Scz but also implies that the regulation of neural excitability by stimulus predictability is awry in the disorder: as one might expect if prior beliefs are less precise. Indeed, reducing prior precision in a hierarchical Bayesian (predictive coding) model attenuates the prediction error responses to oddballs [5].

There are also differences between Scz and controls' resting-state functional magnetic resonance imaging (rsfMRI) responses. Scz show greater power and variability of cortical rsfMRI data, especially in association cortices [8], and also greater connectivity (i.e. rsfMRI data correlations) between association areas [9]. Models of interacting cortical areas producing rsfMRI data can reproduce these effects if E/I balance within cortical areas is increased, although increasing coupling between areas also has similar effects [9]: it is hard to distinguish these model perturbations using fMRI data, as it is less temporally precise.

The most complete modelling approach is to relate neural function to behaviour using the same model. This is a complex procedure and there are few examples in Scz research. One successful example used a spiking network model consisting of pyramidal cells and interneurons to predict spatial working memory performance under ketamine or placebo [10]. Increasing E/I balance in this network (as ketamine is thought to do) allows activity to spread laterally through the network over time, making the spatial 'memory' less precise and predicting increased false alarms to nearby nontarget probes in a spatial working memory task, as is seen under ketamine and also in Scz [11]. The persistent neural activity in the spiking spatial working memory model takes the form of a 'bump attractor', i.e. a subset of neurons which sustain activity from an input over time (the bump) whilst inhibiting local spread of this activity via inhibitory interneurons.

Attractors are essentially quasi-stable states of neural firing that can be implemented in a variety of ways. The first 'attractor networks' were designed to model the storage (and reactivation) of memories in patterns of synaptic weights [12]. In such networks, firing patterns more easily shift towards 'low energy' states, in which strongly connected neurons are active and other neural activity is low. Once in such a state, the network has to receive a large perturbation to shift its firing pattern into a different state. If the energy of the network is plotted as a function of the neural firing patterns, one can visualise these low energy states as 'basins' in an energy landscape. The deeper the basin, the more difficult it is for the network to be shifted out of it. As well as modelling mnemonic processes, similar networks can also perform decision-making [13] and Bayesian belief updating [14].

For more than a decade, it has been hypothesised that changes in neural function in Scz might reduce the stability of cortical attractor states [15]. In particular, NMDAR dysfunction on both recurrent synapses on to pyramidal cells and on inhibitory interneurons could make firing patterns harder to sustain over time and less able to inhibit other firing patterns (respectively), making attractor basins more shallow. In this case, it would be more easy for the network to shift from one state to another – either due to an input that favours the other state or just to random neuronal spiking – but hard to maintain or 'deepen' any one state (Fig. 8.1).

A loss of stable neural states was recently demonstrated in visual cortex of two animal models of schizophrenia [16], and interestingly, healthy volunteers given ketamine (which blocks NMDARs and is used as a model of psychosis) show a decrement in updating to consistent stimulus associations and also increased decision stochasticity in this context [17]. In the remainder of this chapter, I shall describe a recent attempt to model alterations in (Bayesian) belief updating in Scz using a computational model designed to mimic the effects on inference of underlying neural attractor states with varying stability.

It has been known for decades that Scz tend to use less evidence than healthy controls to make decisions in belief updating tasks. The paradigm used to demonstrate this effect is often some variation of the 'beads' or 'urn' task [18], in which the participant is shown two jars containing beads of two colours in opposing ratios

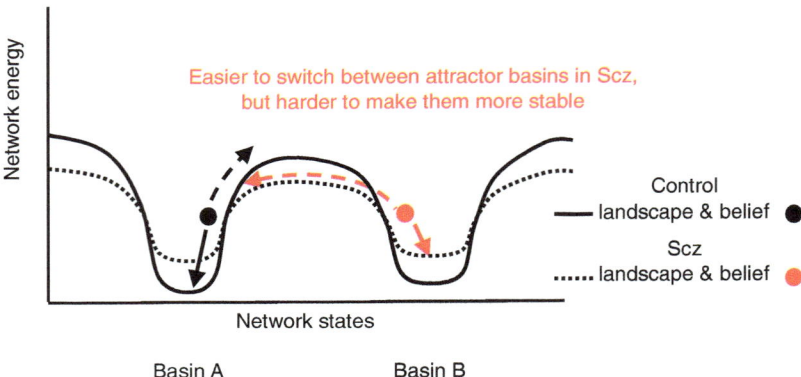

Fig. 8.1 Potential effects of attractor network dynamics on belief updating

(e.g. 80:20 and 20:80 ratios of red/blue beads). The jars are then concealed and a sequence of beads drawn (with replacement) from one jar, and the participant has to either stop the sequence when they are sure of the source jar or give a probability estimate of either jar being the source for the entire sequence. The former version is known as the 'draws to decision' task and the latter as the 'probability estimates' task.

Well-replicated findings in the beads task include many Scz deciding on the jar identity after seeing only one or two beads [19] – the so-called 'jumping to conclusions' bias – and also Scz adjusting their beliefs more than controls after seeing unexpected evidence, termed as 'disconfirmatory bias' [18, 20–23]. Although these biases appear to involve greater belief updating (i.e. higher learning rates) in Scz than in controls, in other tasks Scz seem to update less than controls – especially to longer sequences of more consistent evidence [24] – and Scz are often more stochastic in their responding [25, 26]. These three effects – greater updating to unexpected evidence, lower updating to consistent evidence, and greater stochasticity – are all consistent with an 'unstable attractor' model of belief updating, in which it is easy to switch from one state into another, but hard to stabilise (increase confidence in) any one state, and in which updates are more vulnerable to stochastic fluctuations in neural firing.

Adams et al. [27] used a hierarchical Bayesian model (the hierarchical Gaussian filter [28] – a variational Bayesian model with individual priors) to model belief updating in two independent 'probability estimates' beads task datasets (Fig. 8.2). Models with a standard learning rate ω and response stochasticity ν or including a parameter increasing updating to 'disconfirmatory evidence' φ or a parameter encoding belief instability κ_1 (Fig. 8.3) were formally compared.

In these models, the belief about the jar on trial $k + 1$, $x_2^{(k+1)}$ evolved according a Gaussian random walk of variance $\exp(\omega)$:

$$p\left(x_2^{(k+1)}\right) \sim \mathcal{N}\left(x_2^{(k)}, \exp(\omega)\right)$$

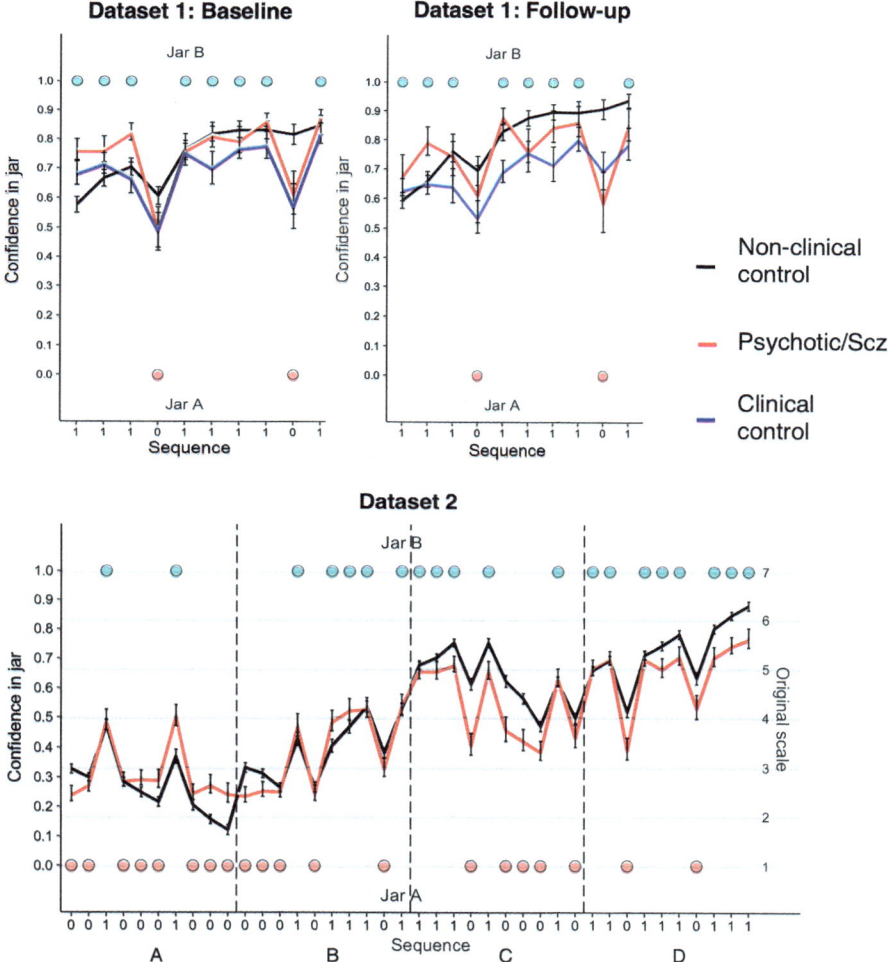

Fig. 8.2 Two beads task datasets. Dataset 1 [23], $n = 80$, including Scz and both clinical and non-clinical controls, tested both when unwell and in recovery, and Dataset 2, $n = 167$, including Scz and non-clinical controls, tested as stable outpatients. In Dataset 2, subjects were each tested on four separate sequences, which are shown concatenated together

In the response model, stochasticity ν determined the width of the beta distribution centred on the current estimate of the jar probability (i.e. the prediction for the next trial), $\hat{\mu}_1^{(k+1)} \equiv s\left(\mu_2^{(k)}\right)$; here μ denotes the current estimate of x, and s is the sigmoid function.

In the 'disconfirmatory bias' model, changes in x_2 from trial to trial occurred according to an autoregressive (AR(1)) process controlled by three parameters: m, the level to which x_2 is attracted; φ, the rate of change of x_2 towards m; and ω, the

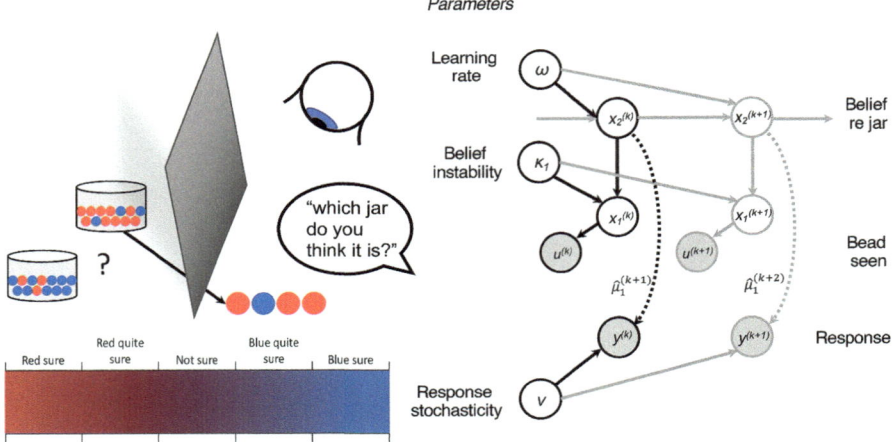

Fig. 8.3 Left – the 'probability estimates' beads task; right – the winning model. The right panel is a schematic representation of the generative model containing belief instability parameter κ_1. The black arrows denote the probabilistic network on trial k; the grey arrows denote the network at other points in time. The perceptual model lies above the dotted arrows and the response model below them. The shaded circles are known quantities, and the parameters and states in unshaded circles are estimated. The dotted line represents the result of an inferential process (the response model builds on a perceptual model inference); the solid lines are generative processes. The response model maps from $\hat{\mu}_1^{(k+1)}$ (purple line) – the probability the blue jar is the source (x_1) on the next trial, itself a sigmoid function of the tendency towards the blue jar (x_2) – to $y^{(k)}$, the subject's indicated estimate of the probability the jar is blue. (See Adams et al. (submitted) for a full description of the model)

variance of the random process:

$$p\left(x_2^{(k+1)}\right) \sim \mathcal{N}\left(x_2^{(k)} + \varphi\left(m - x_2^{(k)}\right), \exp(\omega)\right)$$

Given there was no bias towards one jar or the other, m was fixed to 0, so φ always acted to shift the model's beliefs back towards maximum uncertainty (i.e. disconfirm the current belief) about the jar.

In the 'belief instability' model, changes in μ_2 from trial to trial occur according to two parameters: ω, the variance of the random process, and κ_1, a scaling factor that changes the size of updates when $\hat{\mu}_1 = 0.5$, or maximum uncertainty, relative to when $\hat{\mu}_1$ is closer to 0 or 1, $\hat{\mu}_1^{(k+1)} \equiv s\left(\mu_2^{(k)}\kappa_1\right)$. The effect of increasing κ_1 was to increase updating to unexpected evidence, but decrease updating to consistent evidence (Fig. 8.4), as might be seen in a more unstable attractor network (although note that this model is merely simulating attractor network properties: it does not contain attractor states).

The model containing learning rate ω, response stochasticity ν, and belief instability κ_1 won in all subjects in both datasets. Scz had greater belief instability

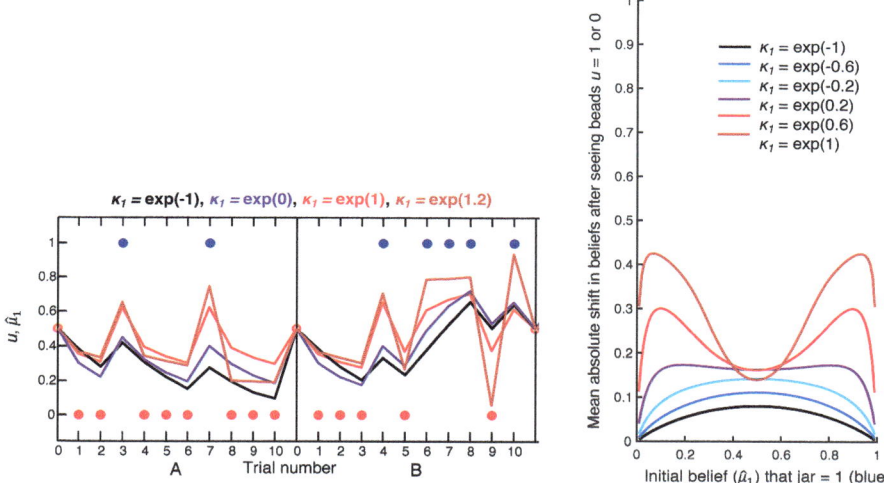

Fig. 8.4 The effects of κ_1 on belief updating. Left panel: simulated belief updating data $\hat{\mu}_1$ (in response to the bead sequence u on the plot) shows that higher κ_1 leads to overweighting of unexpected evidence (instability) and underweighting of consistent evidence. Right panel: this plot illustrates the average absolute shifts in beliefs on observing beads of either colour. This 'vulnerability to updating' is analogous to the 'energy state' of a neural network model (schematically illustrated in Fig. 8.1) – i.e. in low energy states, less updating is expected. The effect of increasing κ_1 is to convert confident beliefs about the jar (near 0 and 1) from low to high 'energy states', i.e. to make them much more unstable

(κ_1) and response stochasticity (ν) than non-clinical controls in both datasets (Fig. 8.5). These parameters correlated in both datasets (Spearman's $\rho = -0.38, -0.52$ and -0.35; all $p < 0.0001$). Interestingly, when unwell, clinical controls' parameter distributions resembled those of Scz; but at follow-up, they resembled non-clinical controls.

Two computational studies of similar tasks in Scz have also demonstrated similar patterns of belief updating. Jardri et al. [29] showed that on average, Scz 'overcount' the likelihood (i.e. the sensory evidence, in Bayesian terms) in a single belief update: the authors attributed this effect to disinhibited cortical message passing, but it could also be due to the belief instability in the model above. Likewise, Stuke et al. [30] showed in another beads task variant that Scz updated more than controls to 'irrelevant information' (i.e. disconfirmatory evidence).

In conclusion, these results show that Scz subjects in two independent beads task datasets have consistent differences in two parameters of a belief updating model that attempts to reproduce consequences of attractor network instability. More detailed spiking network modelling, pharmacological (or other NMDAR) manipulations, and imaging are required in the future to understand how neuromodulatory function in both pyramidal cells and inhibitory interneurons contributes to attractor dynamics and probabilistic inference.

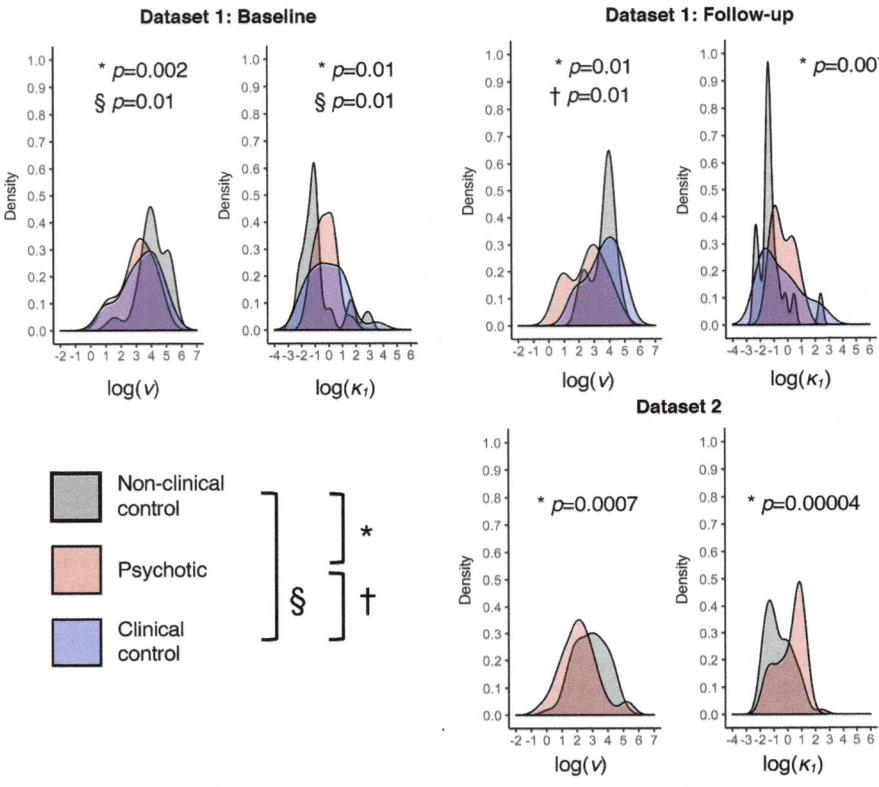

Fig. 8.5 Parameter differences between Scz and clinical and non-clinical controls. Scz consistently had higher belief instability κ_1 and greater response stochasticity ν than non-clinical controls; clinical controls resembled Scz when unwell and non-clinical controls when better

References

1. Harrison PJ (2015) Recent genetic findings in schizophrenia and their therapeutic relevance. J Psychopharmacol Oxf Engl 29:85–96
2. Javitt DC, Zukin SR, Heresco-Levy U, Umbricht D (2012) Has an angel shown the way? Etiological and therapeutic implications of the PCP/NMDA model of schizophrenia. Schizophr Bull 38:958–966
3. Paoletti P, Bellone C, Zhou Q (2013) NMDA receptor subunit diversity: impact on receptor properties, synaptic plasticity and disease. Nat Rev Neurosci 14:383–400
4. Weickert CS et al (2012) Molecular evidence of N-methyl-D-aspartate receptor hypofunction in schizophrenia. Mol Psychiatry. https://doi.org/10.1038/mp.2012.137
5. Adams RA, Stephan KE, Brown HR, Frith CD, Friston KJ (2013) The computational anatomy of psychosis. Front Psychiatry 4:47
6. Umbricht D, Krljes S (2005) Mismatch negativity in schizophrenia: a meta-analysis. Schizophr Res 76:1–23
7. Ranlund S et al (2016) Impaired prefrontal synaptic gain in people with psychosis and their relatives during the mismatch negativity. Hum Brain Mapp 37:351–365

8. Yang GJ et al (2014) Altered global brain signal in schizophrenia. Proc Natl Acad Sci U S A 111:7438–7443
9. Yang GJ et al (2016) Functional hierarchy underlies preferential connectivity disturbances in schizophrenia. Proc Natl Acad Sci U S A 113:E219–E228
10. Murray JD et al (2014) Linking microcircuit dysfunction to cognitive impairment: effects of disinhibition associated with schizophrenia in a cortical working memory model. Cereb Cortex N Y N 1991 24:859–872
11. Mayer JS, Park S (2012) Working memory encoding and false memory in schizophrenia and bipolar disorder in a spatial delayed response task. J Abnorm Psychol 121:784–794
12. Hopfield JJ (1982) Neural networks and physical systems with emergent collective computational abilities. Proc Natl Acad Sci U S A 79:2554–2558
13. Wang X-J (2013) The prefrontal cortex as a quintessential 'cognitive-type' neural circuit: working memory and decision making
14. Gepperth A, Lefort M (2016) Learning to be attractive: probabilistic computation with dynamic attractor networks. In: 2016 joint IEEE international conference on development and learning and epigenetic robotics (ICDL-EpiRob). p 270–277. https://doi.org/10.1109/DEVLRN.2016.7846831
15. Rolls ET, Loh M, Deco G, Winterer G (2008) Computational models of schizophrenia and dopamine modulation in the prefrontal cortex. Nat Rev Neurosci 9:696–709
16. Hamm JP, Peterka DS, Gogos JA, Yuste R (2017) Altered cortical ensembles in mouse models of schizophrenia. Neuron 94:153–167.e8
17. Vinckier F et al (2016) Confidence and psychosis: a neuro-computational account of contingency learning disruption by NMDA blockade. Mol Psychiatry 21:946–955
18. Garety PA, Hemsley DR, Wessely S (1991) Reasoning in deluded schizophrenic and paranoid patients. Biases in performance on a probabilistic inference task. J Nerv Ment Dis 179:194–201
19. Dudley R, Taylor P, Wickham S, Hutton P (2016) Psychosis, delusions and the 'Jumping to Conclusions' reasoning bias: a systematic review and meta-analysis. Schizophr Bull 42:652–665
20. Langdon R, Ward PB, Coltheart M (2010) Reasoning anomalies associated with delusions in schizophrenia. Schizophr Bull 36:321–330
21. Fear CF, Healy D (1997) Probabilistic reasoning in obsessive-compulsive and delusional disorders. Psychol Med 27:199–208
22. Young HF, Bentall RP (1997) Probabilistic reasoning in deluded, depressed and normal subjects: effects of task difficulty and meaningful versus non-meaningful material. Psychol Med 27:455–465
23. Peters E, Garety P (2006) Cognitive functioning in delusions: a longitudinal analysis. Behav Res Ther 44:481–514
24. Averbeck BB, Evans S, Chouhan V, Bristow E, Shergill SS (2010) Probabilistic learning and inference in schizophrenia. Schizophr Res. https://doi.org/10.1016/j.schres.2010.08.009
25. Moutoussis M, Bentall RP, El-Deredy W, Dayan P (2011) Bayesian modelling of Jumping-to-Conclusions bias in delusional patients. Cogn Neuropsychiatry 16:422–447
26. Schlagenhauf F et al (2013) Striatal dysfunction during reversal learning in unmedicated schizophrenia patients. NeuroImage. https://doi.org/10.1016/j.neuroimage.2013.11.034
27. Adams RA, Napier G, Roiser JP, Mathys C, Gilleen J (2018) Attractor-like dynamics in belief updating in schizophrenia. J Neurosci 38:9471–9485
28. Mathys C, Daunizeau J, Friston KJ, Stephan KE (2011) A Bayesian foundation for individual learning under uncertainty. Front Hum Neurosci 5:39
29. Jardri R, Duverne S, Litvinova AS, Denève S (2017) Experimental evidence for circular inference in schizophrenia. Nat Commun 8:14218
30. Stuke H, Stuke H, Weilnhammer VA, Schmack K (2017) Psychotic experiences and overhasty inferences are related to maladaptive learning. PLoS Comput Biol 13:e1005328

Chapter 9
Modeling Cognitive Processing of Healthy Controls and Obsessive-Compulsive Disorder Subjects in the Antisaccade Task

Vassilis Cutsuridis

Abstract Antisaccade performance deficits in obsessive-compulsive disorder (OCD) include increased error rates and antisaccade latencies. These deficits are generally thought to be due to an impaired inhibitory process failing to suppress the erroneous response. The superior colliculus has been suggested as one of the loci of this impaired inhibitory process. Previously recorded antisaccade performance of healthy and OCD subjects is reanalyzed to show greater variability in mean latency and variance of corrected antisaccades as well as in shape of antisaccade and corrected antisaccade latency distributions and increased error rates of OCD patients compared to healthy controls. A neural accumulator model of the superior colliculus is then employed to uncover the biophysical mechanisms giving rise to the observed OCD deficits. The model shows that (i) the increased variability in latency distributions of OCD patients is due to a more noisy accumulation of information by both correct and erroneous decision signals, (ii) OCD patients are *less* confident about their decisions than healthy controls, and (iii) competition via lateral inhibition between the correct and erroneous decision processes, and *not* a third independent inhibitory signal of the erroneous response, accounts for the antisaccade performance of healthy controls and OCD patients.

Keywords Antisaccade paradigm · Eye movements · Superior colliculus · Accumulator model with lateral inhibition · Response inhibition · Impulse control

9.1 Introduction

In the antisaccade paradigm, participants suppress a reflexive saccade (error prosaccade) in favor of a saccade to a position in the opposite hemifield (correct

V. Cutsuridis (✉)
School of Computer Science, University of Lincoln, Lincoln, UK
e-mail: vcutsuridis@lincoln.ac.uk

© Springer Nature Switzerland AG 2019
V. Cutsuridis (ed.), *Multiscale Models of Brain Disorders*, Springer Series in
Cognitive and Neural Systems 13, https://doi.org/10.1007/978-3-030-18830-6_9

antisaccade) [16]. Two processes take place during this paradigm: (1) suppression (or inhibition) of an error prosaccade toward the peripheral stimulus and (2) generation of a volitional saccade to the opposite direction (antisaccade) [14, 20]. The reaction times (RT) of error prosaccades, antisaccades, and corrected antisaccades and the error rate are some of the measures of antisaccade performance [17] with the error rate being the most reliable measure of it. A large study of healthy young males has reported that error prosaccade and antisaccade RTs are highly variable and the error rate is about 20–25% [12, 29].

A recent experimental study reported an increase in error rates and in latency of corrected antisaccades in OCD patients [12]. The antisaccade performance deficit in OCD was speculated to be due a common dysfunctional network of brain structures including the (pre)frontal and posterior parietal cortices and superior colliculus (SC). In this network there is a reported deficit in erroneous response inhibition control [5].

Models of decision-making involve a gradual accumulation of information concerning the various potential responses [6, 7, 9, 10, 11, 23, 24]. As soon as the target appears, a decision process starting at some baseline level T_0 representing the prior expectation begins to rise at a constant rate r until it reaches a threshold T_h representing the confidence level required before the commitment to a particular course of action. Once T_h is crossed, then a response toward the target is initiated. Response time (RT) is the time from the onset of the decision process till when the decision signal crosses T_h. The rate of rise is sometimes assumed to vary randomly from trial to trial, with a mean μ and variance σ^2 [28]. Changes in the baseline level of activity, the rate of rise, or the threshold often result in changes in response latency. Prior expectation and level of activation of intention influence the baseline levels of activation.

The scope of the present modeling study is to uncover what goes wrong neurally (i.e., the mechanisms) in OCD, so the model's behavior best fits the experimental observations (error rates and response time distributions) from both participant populations (healthy controls and OCD patients). For this reason previously recorded error rates and latencies of healthy and OCD participants [12, 13] were reanalyzed to show that OCD patients display higher error rates, increases in mean latency and variance of corrected antisaccades, and greater variability in shape of antisaccade and corrected antisaccade latency distributions relative to healthy participants. The Cutsuridis and colleagues [10] model was then employed to decipher the biophysical mechanisms that gave rise to these antisaccade performance deficits in OCD. The model showed that (i) increased variability in latency distributions of OCD patients was due to a more noisy accumulation of information by both (pre)frontal and posterior parietal centers representing the volitional (correct antisaccade) and reactive (erroneous prosaccade) decision signals, respectively, (ii) OCD patients were *less confident* about their decisions compared to healthy controls (i.e., the decision threshold level T_h value is lower in healthy controls than in OCD patients), and (iii) competition between the correct and erroneous decision processes, and *not* a third top-down STOP of the erroneous response, accounted for the antisaccade performance of both healthy controls and OCD patients.

9.2   Experimental Data

The data (antisaccade performance of healthy controls and OCD patients) were derived from two previously published studies [12, 13]. Detailed description of these data including details about the participants, eye movement recordings, task description, and analysis can be found in Cutsuridis [8] study.

9.3   The Model

The model with its mathematical formalism was initially introduced in Cutsuridis et al. [10] study. Interested readers are referred to this study for detailed description of the model. To assist the readers of this chapter and increase the readability of it, a brief description of the model is provided here. The model is a one-layer SC neural network with lateral inhibition and firing rate nodes (neurons) representing the SC buildup neurons (Fig. 9.1a). The total number of nodes in the network is N. Short-range lateral excitation and long-distance lateral inhibition are assumed between all nodes in model. The lateral interaction kernel w_{ij}, which allows for lateral interactions between model nodes, is a shifted Gaussian, which depends only on the spatial distance between nodes, and it is positive for nearby nodes to the node activated by the input and negative for distant nodes (Fig. 9.1b). Model inputs are of two types: (1) a reactive input (I_r), which represents the error prosaccade decision signal and is hypothesized to originate from the posterior parietal cortices [20], and (2) a planned input (I_p), which represents the correct antisaccade decision signal and originates in the model from the frontal cortical areas [20]. Each input is integrated in opposite model half according to the following way: if the reactive input activates a node and two of each nearest neighbors on each side in the left model half, then the planned input activates the mirror node and its two nearest neighbor nodes on each side in the right model half and vice versa. The strengths of the external inputs are not equal ($I_p > I_r$; see Table 9.3 for values). The reactive input is presented first at time $t = 50$ ms, followed by the planned input, which is presented 50 ms later ($t = 100$ ms) in accordance to experimental evidence [2]. Both inputs remain active for 600 ms.

9.4   Results

9.4.1   Experimental Latency Distributions

The controls and OCD patient experimental data [12, 13] are reanalyzed here using the methodology presented in Cutsuridis and colleagues [10] study. The mean inter-individual of the median intraindividual RT for the error prosaccades was found to

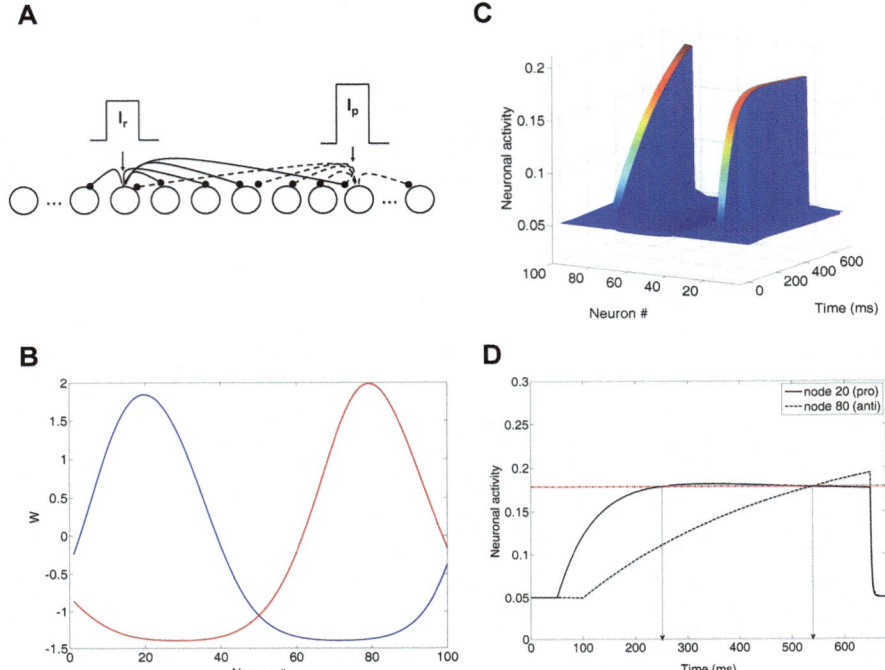

Fig. 9.1 Adapted with permission from Cutsuridis et al. [10]. (**a**) Neural network model. Neurons are represented as firing rate nodes. Short-range lateral excitation and long-distance lateral inhibition were assumed between all nodes in the network. The left model half was activated by a reactive input I_r representing the error prosaccade decision signal, whereas the right model half was activated by a planned input I_p representing the antisaccade decision signal. The strengths of the inputs were not equal ($I_p = 1.5*I_r$). (**b**) Lateral interaction kernels W for nodes 20 and 80 modeled as a shifted Gaussians. The kernels for nodes 20 and 80 were excitatory for the nearby nodes and inhibitory for the distant ones. (**c**) Neuronal activities of all nodes in the network as a function of time (ms). (**d**) Neuronal activity of nodes 20 and 80 as a function of time. Node 20 encoded the reactive input (error prosaccade), and node 80 encoded the planned input (antisaccade). When both activities crossed the threshold (dotted horizontal line), then an eye movement decision was made. In this case, an error prosaccade was initiated first followed by a corrected antisaccade

be 211.09 ms (SD, 49.71) for the controls and 203.81 ms (SD, 53.17) for the patients (Fig. 9.2a; see Table 9.1). This 7.28 ms difference was not statistically significant ($t_{64} = 0.57$, $P = 0.57$). The RT distributions for patients were not much broader than those for the controls, indicating a smaller RT variability. The group coefficient of variation of RT defined as the interquartile RT range ($Q_{75} - Q_{25}$) divided by the median RT was not significantly different for the patients (0.35; SD, 0.21) and the controls (0.30; SD, 0.21) ($t_{64} = 0.4$, $P = 0.85$) (see Table 9.2).

An average cumulative RT distribution for each group (controls vs. patients) (Fig. 9.3a) was computed to further investigate if there is a shape difference between the controls and patients distributions by organizing the RT for each subject in

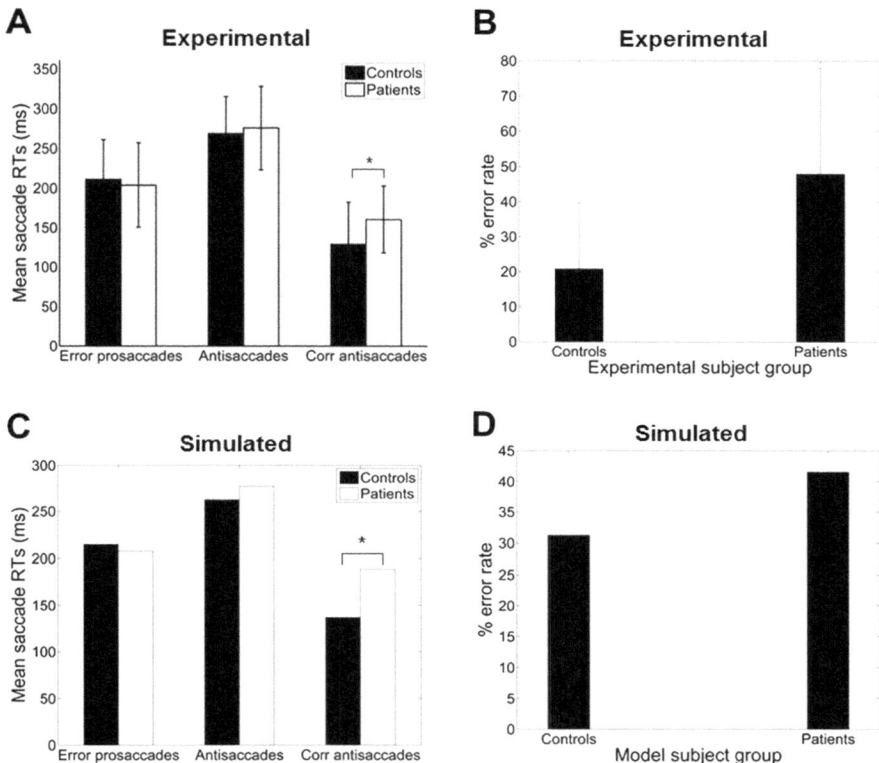

Fig. 9.2 (**a**) Mean of median error prosaccade, antisaccade, and corrected antisaccade reaction times (RTs) for controls and OCD patients. (**b**) Mean percent error rate of controls and OCD patients performing the antisaccade task. (**c**) Simulated median error prosaccade, antisaccade, and corrected antisaccade reaction times (RTs) for controls and OCD patients. (**d**) Simulated percent error rate for controls and OCD patients performing the antisaccade task

Table 9.1 Simulated and experimental median saccade reaction times and their standard deviations and percent error rates for controls and patients with OCD

	Median RT in ms						% error rate	
	Error prosaccade		Antisaccade		Corrected antisaccade			
	Sim	Exp	Sim	Exp	Sim	Exp	Sim	Exp
Controls	214.72	211.09 ± 49.71	262.72	268.61 ± 46.76	136.97	128.84 ± 53.62	31.24	20.79 ± 0.19
OCD patients	207.84	203.81 ± 53.17	277.58	275.73 ± 52.68	188.92	160.34 ± 42.55	41.58	47.96 ± 0.3

ascending order and percentile values were calculated (e.g., the RT for the 5% percentile, the 10% percentile, the 15% percentile, ..., the 95% percentile, the 100% percentile). The percentile values were then averaged across the group to

Table 9.2 Simulated and experimental coefficients of variation (CV) of error prosaccades, antisac-cades, and corrected antisaccades for controls and patients with OCD performing the antisaccade task

	Coefficient of variation (CV)					
	Error prosaccade		Antisaccade		Corrected antisaccade	
	Simulated	Experimental	Simulated	Experimental	Simulated	Experimental
Controls	0.22	0.30 ± 0.21	0.19	0.24 ± 0.07	0.77	0.83 ± 0.41
OCD patients	0.32	0.35 ± 0.21	0.26	0.31 ± 0.12	0.77	0.54 ± 0.24

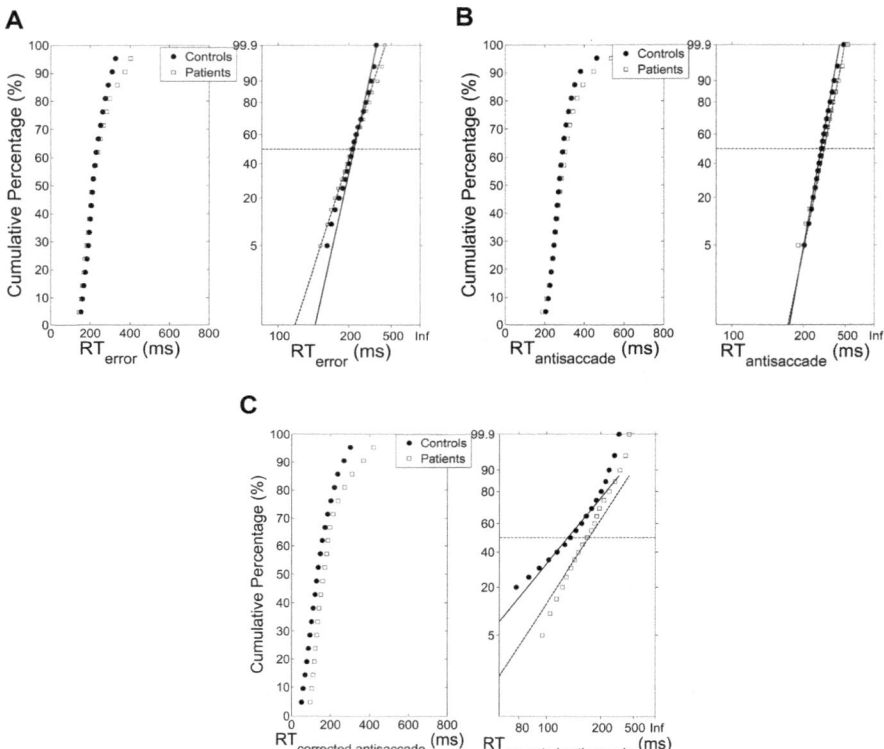

Fig. 9.3 (Left) Average cumulative RT distribution for controls (white empty circles) and patients (black squares). (Right) Reciprobit plots of the average cumulative RT distributions. The x-axis represents 1/RT, and it has been reversed so that RTs increase to the right. Instead of 1/RT values, the axis is marked with the corresponding RT values. The fitted lines correspond to linear regression on the data of each distribution (controls vs. patients). (**a**) Error prosaccades. (**b**) Antisaccades. (**c**) Corrected antisaccades

give average group percentile values. It has been shown that the average distribution retains the basic shape characteristics of the individual distributions [26]. To test the difference between the group distributions for patients and controls, a Wilcoxon rank sum test was used. It can be observed that the two cumulative distributions did not differ in shape ($Z = 1.008$, $P = 0.31$).

A similar analysis was used for the antisaccades and corrected antisaccades for both the controls and patients. The mean inter-individual of the median intraindividual RT for the antisaccades was 268.61 ms (SD, 46.76 ms) for the controls and 275.73 ms (SD, 52.68 ms) for the patients (Fig. 9.2a; see Table 9.1). This 7.12 ms difference was not statistically significant ($t_{64} = 0.57$, $P = 0.57$). The coefficient of variation of antisaccade RTs was also not significantly different for the patients (0.31; SD, 0.12) and for the controls (0.24; SD, 0.07) ($t_{64} = 2.62$, $P = 0.31$) (see Table 9.2).

The average cumulative RT distribution for each group (controls vs. patients) (Fig. 9.3b) was computed as before. To test the difference between the antisaccade group distributions for patients and controls, a Wilcoxon rank sum test was used. It can be observed that the two cumulative distributions differ in shape, and this difference was significant ($Z = 2.65$, $P = 0.008$).

The mean inter-individual of the median intraindividual RT for the corrected antisaccades was 128.84 ms (SD, 53.62 ms) for the controls and 160.34 ms (SD, 42.55 ms) for the patients (Fig. 9.2a; see Table 9.1). This 31.5 ms difference was statistically significant ($t_{64} = 2.60$, $P = 0.0115$). The coefficient of variation of RT was found to be significantly different for the controls (0.83; SD, 0.41) and the patients (0.54; SD, 0.24) ($t_{64} = 3.42$, $P = 0.0011$) (see Table 9.2).

The average cumulative RT distribution for each group (controls vs. patients) (Fig. 9.3c) was similarly computed, and a Wilcoxon rank sum test was used to test the difference between the group distributions for patients and controls. The two cumulative distributions differed in shape, and this difference was significant ($Z = 3.92$, $P < 10^{-3}$).

9.4.2 Simulated Latency Distributions

To simulate the experimental data, the Cutsuridis and colleagues neural network model of the antisaccade performance of healthy controls and schizophrenia patients [10] was employed and extended it into the realm of OCD (Fig. 9.1a). To fit the experimental data, in each trial run in the left and right SC, the time constants τ of the internal states of each node took values from two different normal distributions with means μ_1 and μ_2 and standard deviations σ_1 and σ_2, respectively. The model was run for 5000 trials. In each trial the error prosaccade, antisaccade, and corrected antisaccade latencies were recorded. In the model the error prosaccade reaction time was estimated as the time interval from the onset of the reactive input until the time the activity of the node encoding the reactive input reached a preset threshold (T_h) plus an additional 30 ms (Fig. 9.1d). The antisaccade reaction time was estimated as the time interval from the onset of the reactive input until the time the activity of the node encoding the planned input reached the threshold plus 30 ms (Fig. 9.1d). The corrected antisaccade reaction time was the time interval from threshold crossing of the error node activity until the threshold crossing of the correct node activity.

To simulate the error prosaccade, antisaccade, and corrected antisaccade RT distributions as well as the error rates of both healthy controls and OCD participant groups, the integration constants τ (μ and σ) for both nodes that integrated the reactive (μ_1 and σ_1) and planned (μ_2 and σ_2) inputs were varied. In the control condition, $\mu_1 = 0.01787$, $\sigma_1 = 0.003$, $\mu_2 = 0.0056$, and $\sigma_2 = 0.0016$, whereas in OCD condition $\mu_1 = 0.0165$, $\sigma_1 = 0.005$, $\mu_2 = 0.0047$, and $\sigma_2 = 0.002$. In both conditions, the threshold value at which as a decision was reached (parameter T_h in Table 9.3) was higher in OCD patients than in healthy controls. The simulated median RTs for the error prosaccades, antisaccades, and corrective antisaccades were 214.72 ms, 262.72 ms, and 136.97 ms, respectively, for the model controls and 207.84 ms, 277.58 ms, and 188.917 ms, respectively, for the model patients. The simulated median RT values are very close to the experimental ones (Fig. 9.2c; see also Table 9.1). The simulated coefficients of variation (CVs) for the error prosaccades, antisaccades, and corrected antisaccades were 0.22, 0.19, and 0.77, respectively, for the controls and 0.32, 0.26, and 0.77, respectively, for the patients. The simulated CV values are very close to the experimental ones (see Table 9.2).

As before the simulated average cumulative RT distributions for error prosaccades, antisaccades, and corrected antisaccades for both groups (model controls vs. model patients) were estimated by organizing the RT for each subject group from each trial run in ascending order and calculating the percentile values (e.g., the RT for the 5% percentile, the 10% percentile, the 15% percentile, ..., the 95% percentile) were computed. The percentile values were then averaged across trial runs (5000 trial runs) for each subject group to give average subject group percentile values. Carpenter and Williams [4] showed that if the cumulative RT distribution is plotted using 1/RT in a reciprobit plot, then the RTs will fall on a straight line. Thus, the average cumulative distribution data of RT (error prosaccade, antisaccade, and corrected antisaccade) for the experimental and simulated controls and patients in a reciprobit plot were transformed. A best-fitting regression line was computed for

Table 9.3 Model parameters and values

| Symbol | Value | | Symbol | Value |
	Controls	OCD		
T_h	0.16	0.177	σ	$2\pi/10$
C	0.35		Δx	$2\pi/N$
I_r	1		A	1
I_p	1.5		N	100
μ_1	0.01787	0.0165	β	0.5
σ_1	0.003	0.005	θ	0.5
μ_2	0.0056	0.0047	μ_n	0
σ_2	0.0016	0.002	σ_n	0.05
T	50 ms, unless mentioned otherwise		ntrials	5000

each behavioral category (error prosaccade, antisaccade, and corrected antisaccade) in each subject group (controls and patients). An R correlation coefficient was estimated to assess how good fit was the modeled regression line to the experimental data. The model fit for each behavioral category and for subject group was excellent (correlation coefficient R was 0.99 for error prosaccades and antisaccades and 0.96 for corrected antisaccades in the healthy control group and 0.99 for error prosaccades and antisaccades and 0.97 for corrected antisaccades in the OCD group). To compare the two simulated regression lines for the patient and control groups, the homogeneity of slope and intercept regression analysis described in Wuensch [33] was used. The coefficients (slope and intercept) were extracted and fitted to the experimental 1/RT data (see right plots of Fig. 9.3a–c). A comparison of the homogeneity of slopes and intercepts showed that both (controls and patients) fitted error prosaccade lines were statistically different in slope ($t_{36} = 5.53305$, $p = 0.005$) and in intercept ($t_{36} = 2.9$, $p = 0.005$). A similar comparison of the slopes and intercepts was made for the antisaccades and corrected antisaccades for the controls and patients. The fitted antisaccade lines were not statistically different in slope ($t_{36} = 2.10387$, $p = 0.005$) and in intercept ($t_{36} = 1.75$, $p = 0.005$). The fitted corrected antisaccade lines were not statistically different in slope ($t_{36} = 2.49$, $p = 0.005$) and in intercept ($t_{36} = 0.193$, $p = 0.005$).

9.4.3 Error Rates

The experimental error rate was found to be 20.79% for the controls and 47.96% for the patients (Fig. 9.2b; see also Table 9.1). In the model an error was considered when the firing activity of the node encoding the reactive input (error prosaccade) crossed a preset threshold level. The model error rate was estimated to be 31.24% for the controls and 41.58% for the patients (Fig. 9.2d; see Table 9.1), thus qualitatively reproducing the increasing error rate trend reported in OCD patients.

9.5 Discussion

9.5.1 What Have We Learned from This Model?

Previously recorded antisaccade performance of healthy and OCD subjects [12] is reanalyzed to show greater variability in mean latency and variance of corrected antisaccades as well as variability in shape of antisaccade and corrected antisaccade latency distributions and increased error rates of OCD patients relative to healthy participants. A well-established neural nonlinear accumulator model of antisaccade performance is then employed to uncover the biophysical mechanisms giving rise to these observed OCD deficits. The major finding of this study is that the brains of OCD participants when they are performing the antisaccade task are noisier than the brains of healthy controls. This noise is reflected in the rate of accumulation of information (μ and σ) and the threshold level S_T (confidence level required before commitment to a particular course of action). As we can see from Table 9.3, the value of T_h (threshold level S_T) is higher in the OCD patient case than in healthy control one meaning that the OCD patients are less confident about their decisions than the healthy controls. Their lack of confidence is reflected in their latencies, which are longer than the control ones (see Table 9.1). Parameters μ_1 and μ_2 (see Table 9.3 for values) are greater in control condition than in OCD condition meaning that error prosaccades, antisaccades, and corrected antisaccades are slower in OCD patients than in healthy controls. Similarly, σ_1 and σ_2 (see Table 9.3 for values) are smaller in healthy control condition than in patient one, which means that error prosaccade, antisaccade, and corrected antisaccade latencies are more variable in OCD patients than in healthy participants. A physiological interpretation of the time constant, τ, and its variability may be variability of NMDA-based rate of evidence integration [11]. Experimental studies have shown that NMDA hypofunction is implicated in neurodegenerative disorders such schizophrenia and OCD [19].

Another important finding of this study is the absence of a third signal, inhibitory in nature, necessary to prevent the error prosaccade from being expressed when the antisaccade reached the threshold first. Such a third inhibitory signal has been speculated to exist by Noorani and Carpenter [22–24] in the form of a "stop-and-restart" mechanism that partially captures the antisaccade performance of healthy participants (see the Cutsuridis ([6], 2017) studies for a critique of Noorani and Carpenter [24] model limitations). Recent experimental evidence has demonstrated that lateral interactions within SC intermediate segment are more suitable for faithfully accumulating subthreshold signals for saccadic decision-making [25]. Another experimental study by Everling and Johnston [15] challenges the idea of a third suppressive/inhibitory influence (STOP signal in the Noorani and Carpenter model) of prefrontal cortical areas on reflexive, erroneous prosaccade generation in the antisaccade paradigm. My study has provided quantitative evidence that such a third inhibitory STOP process is not necessary, but instead competition via lateral inhibition in the SC between the neurons encoding the erroneous response (error prosaccade) and neurons encoding the voluntary one (antisaccade) is sufficient to prevent in some trials the error prosaccade from crossing the threshold when the antisaccade has reached it first. My model simulated accurately the latency distributions of the error prosaccades, antisaccades, and corrected antisaccades of both healthy controls and OCD patients.

It has been suggested that when data are plotted on the reciprobit plot, then the resulting straight line on the reciprobit plot could be used a diagnostic tool to assess the contribution of different factors influencing the experimental results [3]. When straight lines swivel by the threshold S_T [28], then the mean and variances of the lines are unequal. When the lines are parallel and shifted by μ, then the slopes ($1/\sigma$) of the lines are equal, but their latency medians are not [27]. When the lines cross, then the slopes are not equal, but their medians are [21]. In my model we observed that when the lines crossed (error prosaccade (Fig. 9.3a) and antisaccade

(Fig. 9.3b)), then the median values of error prosaccade and antisaccade latencies are not significantly equal. When the lines are parallel and shifted (corrected antisaccades; Fig. 9.3c), then the median latencies are significantly different.

9.5.2 Comparison with Other Models

Neurodegenerative disorders such as schizophrenia, bipolar disorder, and major depression share same dimensions of clinical symptoms and same genetic vulnerabilities [1]. OCD, however, does not share with them the same clinical and genetic vulnerability factors [1]. On the other hand, both OCD and schizophrenia have been associated with dysfunctions of similar cortical and subcortical circuits [32]. These dissimilarities are reflected in the antisaccade performances of patients with schizophrenia and those with OCD. Patients with schizophrenia consistently report increased error rate, increased *both* antisaccade and corrected antisaccade latencies, while their erroneous prosaccade ones are not significantly different from those of healthy controls [18, 30, 31]. On the other end, OCD patients show increases in error rates, increases in latencies of *just* the corrected antisaccades, and significant differences in the shapes of OCD latency distributions of antisaccades and corrected antisaccades compared with those of the healthy controls ([12]; this study).

The computational study of Cutsuridis and colleagues [10] showed that the differences in the antisaccade performance of healthy controls and schizophrenia patients are due to a more noisy accumulation of information process (μ and σ) in both frontal (voluntary; antisaccade) and posterior parietal (reactive; erroneous prosaccade) decision centers, but both groups' prior confidence level (decision threshold level S_T) required before commitment to a particular course of action was not affected by disease (schizophrenia). In the present computational study of antisaccade performance of healthy controls and OCD patients, the accumulation of information process (μ and σ) in both frontal (voluntary; antisaccade) and posterior parietal (reactive; erroneous prosaccade) decision centers is still noisy compared to healthy controls, but the OCD patients' confidence level value (decision threshold level S_T) is higher than that of the healthy controls. This means that the OCD patients are less confident to respond to that of the healthy controls. The difference in the confidence level value between schizophrenia (see Table 9.3 in [10]) and OCD (Table 9.3 in this study) participant groups is *maybe* due to the accuracy constraints of the mirror antisaccade task reported in the Cutsuridis and colleagues [10] study making the schizophrenia patients less confident (more hesitant) to respond, which is reflected in the observed increases in their latencies (compare latency values in Table 9.1 in Cutsuridis et al. [10] study and this one).

9.6 Conclusion

Overall, the model showed in a quantitative way why the antisaccade performance of patients with OCD is so poor, that this performance is not due to a deficit in the top-down inhibitory control of the erroneous response as many speculated, but instead it is a product of a neuronal competition via lateral inhibition between the erroneous prosaccade and the antisaccade. The model accurately reproduced the error rates, the median antisaccade, median error prosaccade, and median corrected antisaccade latencies as well as the antisaccade, error prosaccade, and corrected antisaccade distributions of healthy controls and OCD patients. The model showed that the experimentally observed antisaccade performance deficits of OCD patients are due to (i) a more noisy accumulation of information by both erroneous and correct decision signals and (ii) a higher confidence level of the OCD patients. The results presented here illustrate the benefits of tightly integrating psychophysical studies with computational neural modeling, because the two methods complement each other and they may provide together a strong basis for hypothesis generation and theory testing regarding the neural basis of decision-making in health and in disease.

References

1. American Psychiatric Association (2013) Schizophrenia spectrum and other psychotic disorders. In: Diagnostic and statistical manual of mental disorders. American Psychiatric Association, Washington, DC, p 87–122
2. Becker W (1989) Metrics. In: Wurtz R, Goldberg M (eds) Neurobiology of saccadic eye movements. Elsevier, New York, pp 12–67
3. Carpenter RHS (1981) Oculomotor procrastination. In: Fisher DF, Monty RA, Senders JW (eds) Eye movements: cognition and visual perception. Lawrence Erlbaum, Hillsdale, pp 237–246
4. Carpenter RHS, Williams MLL (1995) Neural computation of log likelihood in the control of saccadic eye movements. Nature 377:59–62
5. Chamberlain SR, Blackwell AD, Fineberg NA, Robbins TW, Sahakian BJ (2005) The neuropsychology of obsessive compulsive disorder: the importance of failures in cognitive and behavioural inhibition as candidate endophenotypic markers. Neurosci Biobehav Rev 29:399–419
6. Cutsuridis V (2015) Neural competition via lateral inhibition between decision processes and not a STOP signal accounts for the antisaccade performance in healthy and schizophrenia subjects. Front Neurosci 9:5. https://doi.org/10.3389/fnins.2015.00005
7. Cutsuridis V (2010) Neural accumulator models of decision making in eye movements. Adv Exp Med Biol 657:61–72
8. Cutsuridis V (2017a) A neural accumulator model of antisaccade performance of healthy controls and obsessive-compulsive disorder patients. In: vanVugt MK, Banks AP, Kennedy WG (eds) Proceedings of the 15th international conference on cognitive modeling. University of Warwick, Coventry, pp 85–90
9. Cutsuridis V (2017b) Behavioral and computational varieties of response inhibition in eye movements. Philos Trans R Soc Lond B 372:20160196
10. Cutsuridis V, Kumari V, Ettinger U (2014) Antisaccade performance in schizophrenia: a neural model of decision making in the superior colliculus. Front Neurosci 8:13. https://doi.org/10.3389/fnins.2014.00013

11. Cutsuridis V, Smyrnis N, Evdokimidis I, Perantonis S (2007) A neural model of decision making by the superior colliculus in an antisaccade task. Neural Netw 20:690–704

12. Damilou A, Apostolakis S, Thrapsanioti E, Theleritis C, Smyrnis N (2016) Shared and distinct oculomotor function deficits in schizophrenia and obsessive compulsive disorder. Psychophysiology 53(6):796–805. https://doi.org/10.1111/psyp.12630

13. Evdokimidis I, Smyrnis N, Constantinidis TS, Stefanis NC, Avramopoulos D, Paximadis C et al (2002) The antisaccade task in a sample of 2006 young men. I. Normal population characteristics. Exp Brain Res 147:45–52

14. Everling S, Fischer B (1998) The antisaccade: a review of basic research and clinical studies. Neuropsychologia 36:885–899

15. Everling S, Johnston K (2013) Control of the superior colliculus by the lateral prefrontal cortex. Philos Trans R Soc Lond Ser B Biol Sci 368:20130068. https://doi.org/10.1098/rstb.2013.0068

16. Hallett PE (1978) Primary and secondary saccades to goals defined by instructions. Vis Res 18:1279–1296

17. Hutton SB, Ettinger U (2006) The antisaccade task as a research tool in psychopathology: a critical review. Psychophysiology 43:302–313

18. Karoumi B, Ventre-Dominey J, Vighetto A, Dalery J, d'Amato T (1998) Saccadic eye movements in schizophrenia patients. Psych Res 77:9–19

19. Lewis DA (2012) Cortical circuit dysfunction and cognitive deficits in schizophrenia-implications for preemptive interventions. Eur J Neurosci 35(12):1871–1878

20. Munoz DP, Everling S (2004) Look away: the antisaccade task and the voluntary control of eye movement. Nat Rev Neurosci 5:218–228

21. Nakahara H, Nakamura K, Hikosaka O (2006) Extended later model can account for trial-by-trial variability of both pre- and post-processes. Neural Netw 19:1027–1046

22. Noorani I, Carpenter RHS (2012) Antisaccades as decisions: LATER model predicts latency distributions and error responses. Eur J Neurosci. https://doi.org/10.1111/ejn.12025

23. Noorani I, Carpenter RHS (2013) Antisaccades as decisions: LATER model predicts latency distributions and error responses. Eur J Neurosci 37:330–338. https://doi.org/10.1111/ejn.12025

24. Noorani I, Carpenter RHS (2014) Re-starting a neural-race: antisaccade correction. Eur J Neurosci 39:159–164. https://doi.org/10.1111/ejn.12396

25. Phongphanphanee P, Marino RA, Kaneda K, Yanagawa Y, Munoz DP, Isa T (2014) Distinct local circuit properties of the superficial and intermediate layers of the rodent superior colliculus. Eur J Neurosci 40:2329–2343. https://doi.org/10.1111/ejn.12579

26. Ratcliff R (1979) Group reaction time distributions and an analysis of distribution statistics. Psychol Bull 86(3):446–461

27. Reddi BAJ, Asrress KN, Carpenter RHS (2003) Accuracy, information and response time in a saccadic decision task. J Neurophys 90:3538–3546

28. Reddi BA, Carpenter RHS (2000) The influence of urgency on decision time. Nat Neurosci 3:827–830

29. Smyrnis N, Evdokimidis I, Stefanis NC, Constantinidis TS, Avramopoulos D, Theleritis C et al (2002) The antisaccade task in a sample of 2006 young men. II. Effects of task parameters. Exp Brain Res 147:53–63

30. Smyrnis N, Karantinos T, Malogiannis I, Theleritis C, Mantas A, Stefanis NC et al (2009) Larger variability of saccadic reaction times in schizophrenia patients. Psychiatry Res 168:129–136

31. Theleritis C, Evdokimidis I, Smyrnis N (2014) Variability in the decision process leading to saccades: a specific marker for schizophrenia? Psychophysiology 51:327–336. https://doi.org/10.1111/psyp.12178

32. Tekin S, Cummings JL (2002) Frontal-subcortical neuronal circuits and clinical neuropsychiatry: an update. J Psychosom Res 53:647–654. https://doi.org/10.1016/S0022-3999(02)00428-2

33. Wuensch KL (2007) Comparing correlation coefficients, slopes and intercepts. http://core.ecu.edu/psyc/wuenschk/docs30/CompareCorrCoeff.pdf

Chapter 10
Simulating Cognitive Deficits
in Parkinson's Disease

Sébastien Hélie and Zahra Sajedinia

Abstract Parkinson's disease (PD) is caused by the accelerated death of dopamine–producing neurons. Numerous studies documenting cognitive deficits of people with PD have revealed impairment in a variety of tasks related to memory, learning, visuospatial skills, and attention. In this chapter, we describe a general approach used to model PD and review three computational models that have been used to simulate cognitive deficits related to PD. The models presentation is followed by a discussion of the role of glia cells and astrocytes in neurodegenerative diseases. We propose that more biologically–realistic computational models of neurodegenerative diseases that include astrocytes may lead to a better understanding and treatment of neurodegenerative diseases in general and PD in particular.

Keywords Parkinson's disease · Dopamine · Astrocytes

10.1 What Is Parkinson's Disease?

Parkinson's disease (PD) is a neurodegenerative disease caused by the accelerated death of dopamine (DA)-producing neurons. Cell loss is predominately found in the ventral tier of the substantia nigra pars compacta (SNpc), with less damage in the dorsal tier [1, 2]. In contrast, normal aging yields substantially less cell loss and in a dorsal–to–ventral pattern. Motor symptoms appear after a loss of 60–70% of SNpc cells and 70–80% of DA levels in the striatum (where SNpc cells send projections) [2, 3]. Motor symptoms include resting tremor, rigidity, bradykinesia, and akinesia.

In addition to motor deficits, PD patients present cognitive symptoms that resemble those observed in patients with damage to the frontal lobes. Numerous studies documenting cognitive deficits of PD patients have revealed impairment in a variety of tasks related to memory, learning, visuospatial skills, and attention

S. Hélie (✉) · Z. Sajedinia
Department of Psychological Science, Purdue University, West Lafayette, IN, USA
e-mail: shelie@purdue.edu; zsajedin@purdue.edu

© Springer Nature Switzerland AG 2019   105
V. Cutsuridis (ed.), *Multiscale Models of Brain Disorders*, Springer Series in
Cognitive and Neural Systems 13, https://doi.org/10.1007/978-3-030-18830-6_10

[4]. Critically, these deficits are not limited to challenging or esoteric laboratory experiments, but they are also present in everyday lives [5]. Thus, a thorough understanding of cognitive deficits in PD is essential for improving the quality of life of people with PD. In this book chapter, we review popular computational models that aim to better understand the progression of cognitive deficits in PD.

10.2   How Is Parkinson's Disease Typically Modeled?

Because PD is characterized by the accelerated death of DA–producing neurons in the SNpc, PD is typically modeled by reducing the amount of DA available in computational models [6]. Hence, in theory, any computational model that includes an explicit role for DA could be tested against data from PD patients. Dopamine projections are predominantly directed at the basal ganglia (BG) and frontal cortex [7], so many computational models of these brain structures include a role for DA. In addition, DA is thought to play an important role in BG–frontal functional connectivity by modulating the efficiency of the connectivity [8]. Computational models of cognitive deficits in PD have used one or both of these approaches (i.e., reduced DA levels and/or reduced BG–frontal connectivity) to simulate Parkinsonian symptoms.

A typical BG–cortical loop is shown in Fig. 10.1. As can be seen, cortex sends excitatory connections to the striatum, where the information splits into two

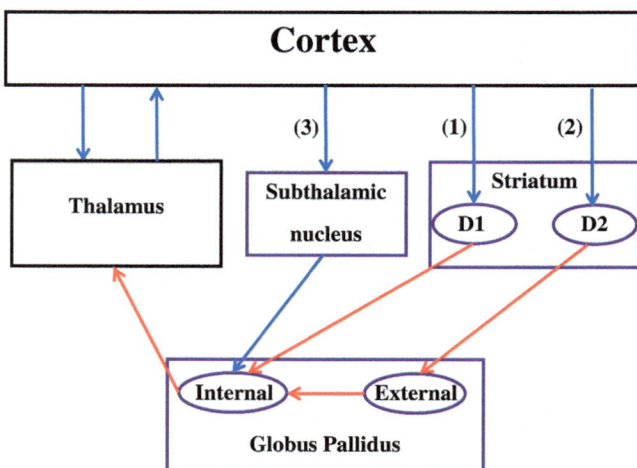

Fig. 10.1 A typical BG–cortical loop. Purple boxes correspond to areas of the BG, while black boxes are not included in the BG. Blue arrows represent excitatory (glutamatergic) connections, while red arrows represent inhibitory (GABA) connections. The direct pathway passes through the D1 dopamine receptors in the striatum, the indirect pathway passes through the D2 dopamine receptors in the striatum, and the hyperdirect pathway passes through the subthalamic nucleus. If the thalamic projections target the same cortical region that initially targeted the striatum, the circuit is called a *closed loop*. Otherwise, the circuit is an *open loop*

pathways: (1) the *direct pathway* and (2) the *indirect pathway*. Most connectivity within the BG is inhibitory, so the overall output of the direct pathway is to excite cortex, whereas the overall output of the indirect pathway is to inhibit cortex. In addition to these two pathways through the striatum, a third pathway called the *hyperdirect pathway* (3) goes through the subthalamic nucleus and directly excites the internal segment of the globus pallidus (GPi). Similar to the indirect pathway, the overall output of the hyperdirect pathway is to inhibit cortex.

10.3   Computational Models of Cognitive Deficits in Parkinson's Disease

In this section, we briefly describe three computational models that have been used to simulate PD in chronological order. The reader is referred to the original papers for implementation details.

10.3.1   Monchi, Taylor, and Dagher (2000)

Monchi and colleagues proposed a model to account for working memory deficits in PD and schizophrenia [9]. The model includes three BG–cortical closed loops: two with the prefrontal cortex (one for spatial information and the other for object information) and one through the anterior cingulate cortex (ACC). The role of the two prefrontal–BG loops is to maintain working memory information about the stimuli, whereas the ACC loop maintains the adopted strategy by inhibiting all the prefrontal cortex loops except one (i.e., representing the selected strategy). In the model, the visual stimulus is input to the prefrontal cortex loops, and the stimulus activity is propagated through the direct pathway of the BG. As a result, the thalamus is released from inhibition of the GPi, and activation produced by the stimulus in the prefrontal cortex reverberates through closed loops with the thalamus. When a response is required, the prefrontal cortex transfers its activation to the premotor cortex. If the response is incorrect, a feedback signal is sent to the ACC loop, which selects a new strategy by switching its inhibition to different prefrontal cortex loops. The Monchi et al. model has been used to simulate a delayed response task and the Wisconsin Card Sorting Test [10]. Interestingly, reducing the connection strengths within the BG–cortical loops produce Parkinsonian symptoms, whereas reducing the strength of the feedback signal produces deficits similar to those observed in schizophrenia [9]. This model thus shows that PD can be simulated by reducing the functional connectivity between the BG and prefrontal cortex, which is consistent one of the roles of tonic DA [6].

10.3.2 Frank (2005)

Frank proposed a model that includes both the direct and indirect pathways through the BG, the premotor cortex, and an unspecified input area [11]. In this model, the input activates both the premotor cortex and the striatum. However, cortical activation is insufficient to produce a response, so BG processing is required to gate the correct response. The focus of the model is on the roles of the indirect pathway and DA in probabilistic learning. In Frank's model, the direct pathway is in charge of selecting the appropriate action (Go), whereas the indirect pathway is in charge of inhibiting inappropriate actions (NoGo). The direct and indirect pathways converge in the GPi and compete to control GPi activation and eventually the response.

The competition between the direct and indirect pathways is modulated by DA. Specifically, higher dopamine levels increase activation in the direct pathway (e.g., through D1 receptors) and reduces activation in the indirect pathway (e.g., through D2 receptors). Hence, tonic DA release following unexpected rewards results in long–term potentiation (LTP) in the direct pathway and long–term depression (LTD) in the indirect pathway. In contrast, DA dips following the unexpected absence of a reward reduces activation and produces LTD in the direct pathway but increases activation and produces LTP in the indirect pathway. The simulation results suggest that the dynamic range of the DA signal is critical in probabilistic learning and reversal learning (e.g., when the response–reward associations are changed during learning). Reducing (to simulate PD) or increasing (to simulate medication overdose) DA levels can result in simulated Parkinsonian symptoms [11]. These results are consistent with the role of tonic DA in reinforcement learning and the possibility of simulating PD symptons by reducing the dynamic range of tonic DA [6].

10.3.3 COVIS

COVIS [12] is a multiple–systems theory that was originally developed to account for the many behavioral dissociations between rule–based and information–integration categorization [13]. COVIS includes a hypothesis–testing system and a procedural learning system. The hypothesis–testing system can quickly learn a small set of categories (i.e., those that can be found by hypothesis–testing), while the procedural learning system can learn any type of arbitrary categories in a slow trial-and-error manner. Each categorization system relies on a separate brain circuit, but they both include the BG. In the hypothesis–testing system, the BG is used to support working memory maintenance and for rule switching. In the procedural learning system, the BG is used to learn stimulus–response associations. While COVIS is the earliest reviewed model in this chapter, it took nearly 15 years before it was used to account for PD [14, 15]. Reducing DA levels in COVIS has been shown to account for many cognitive symptoms in PD patients such as perseveration, category learning deficits (both deterministic and probabilistic), deficits in rule maintenance, and reduced sensitivity to negative feedback (see [14, 15]). Because of its multiple-systems architecture, COVIS can account for both the effect of reducing tonic DA (e.g., perseveration) and phasic DA (e.g., reinforcement learning) [6].

10.4 The Role of Astrocytes in Neurodegenerative Diseases

Recent findings in neurophysiology show that glia astrocytes are doing more than just passive and housekeeping functions in the brain: they also actively modulate synapses and interfere with neural signaling [16–20]. Furthermore, multiple studies show that astrocytes are involved in both the initiation and progression of neurode-generative diseases [21–24]. For example, it has been shown that the number of astrocytes increases in the PD [21, 22], and because astrocytes interfere with neural signaling, the larger number of astrocytes should result in abnormal firing rates in PD-related neural networks. Physiological studies have shown this abnormal firing rate as an increase in neural bursting [25].

The effect of astrocytes on the firing rate of neural networks suggests using models of neural networks for PD that include astrocytes and present a different neural firing rate in comparison to typical neural networks. However, all the reviewed models have focused only on neurons (mostly using rate models), and they did not consider the changes in neural signaling caused by astrocytes. Therefore, to achieve a more accurate biological model of PD, it can be useful to include a model of astrocytes in the simulated neural networks. Note that we are not claiming that using spiking models is superior to using the rate models that were described in Sect. 10.3. Instead, we are claiming that spiking models could provide complementary information that can be useful in better understanding PD at the neural level and possibly address a different set of questions.

The current problem with computational models of astrocytes is that unlike neurons, which have well–developed and widely used computational models [26, 27], computational models of astrocytes have not yet received much attention. To date, a few computational models of astrocytes have been proposed [28–33], but no astrocyte model is fully functional or widely accepted.

To address the lack of biological–plausibility of astrocyte models, we [34] recently proposed a new dynamic model of astrocytes based on the Izhikevich model of neurons [26]. The parametrized model is described by:

$$6\dot{v} = 2.77 \times 10^{-5}(v + 70)(v - 1.43 \times 10^{3}) - u + I \tag{10.1}$$

$$\dot{u} = 0.03\{-6.5 \times 10^{-4}(v + 70) - u\} \tag{10.2}$$

$$if \quad v \geq 35, \quad then \quad v \leftarrow -50, \quad u \leftarrow u + 100 \tag{10.3}$$

where v is the membrane potential of the astrocyte and u is an abstract term representing cell recovery. Unlike previous astrocyte models, Eq. 10.3 shows a linear

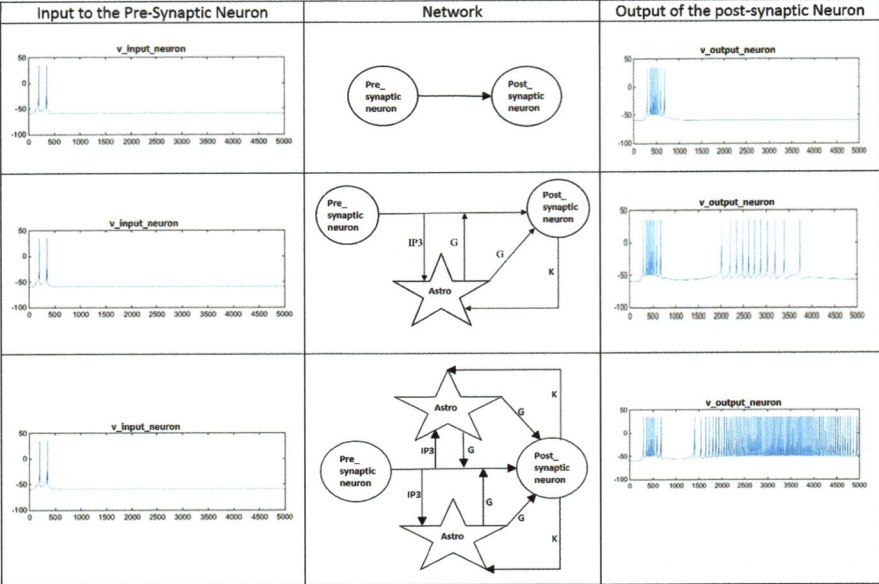

Fig. 10.2 Effect of inclusion of astrocytes in a simple neural network

current/voltage relationship and does not spike, which is in line with biological astrocytes [35]. The astrocyte model has been added in simple spiking neural networks (Fig. 10.2), and the results show that adding an astrocyte increases the length of postsynaptic activity, and that additional astrocytes produce extended bursting behavior, which is compatible with the physiological observations of neuron's outputs in PD [25]. This new astrocyte model has yet to be used in more complex networks to simulate PD symptoms, but the results have been encouraging, and the availability of the model should facilitate future work. The next step in simulating PD symptoms is to add more astrocytes in a simulation model of a cognitive function and observe the resulting behavior.

10.5   Conclusion

This chapter reviewed three computational models that have been used to reproduce cognitive symptoms related to PD. As mentioned in Sect. 10.2, most of these models focus on the BG and their interaction with prefrontal cortex. These brain structures have received much attention from computational modelers. In particular, some computational models of the BG are very anatomically detailed [36, 37], and in theory any computational model of the prefrontal cortex or the BG that includes a role for DA could be used to account for PD. However, as argued in Sect. 10.4, none of these models include glia cells, which have been shown to play an important role in neurodegenerative diseases. A new dynamical model of astrocyte activation has recently been proposed [34], which should facilitate computational work with glia cells. These new developments are exciting, and future computational modeling of neural diseases could lead to important breakthroughs and treatments.

Acknowledgements This research was funded in part by the National Institute of Mental Health, award #2R01MH063760.

References

1. Fearnley JM, Lees AJ (1991) Ageing and Parkinson's disease: substantia nigra regional selectivity. Brain 114:2283–2301
2. Gibb WR, Lees AJ (1991) Anatomy, pigmentation, ventral and dorsal subpopulations of the substantia nigra, and differential cell death in Parkinson's disease. J Neurol Neurosurg Psychiatry 54:388–396
3. Bernheimer H, Birkmayer W, Hornykiewicz O, Jellinger K, Seitelberger F (1973) Brain dopamine and the syndromes of Parkinson and Huntington clinical, morphological and neurochemical correlations. J Neurol Sci 20:415–455
4. Gotham AM, Brown RG, Marsden CD (1988) "Frontal" cognitive function in patients with Parkinson's disease "on"and "off" levodopa. Brain 111:299–321
5. Poliakoff E, Smith-Spark JH (2008) Everyday cognitive failures and memory problems in Parkinson's patients without dementia. Brain Cogn 67:340–350
6. Hélie S, Chakravarthy S, Moustafa AA (2013) Exploring the cognitive and motor functions of the basal ganglia: an integrative review of computational cognitive neuroscience models. Front Comput Neurosci 7:174. http://journal.frontiersin.org/article/10.3389/fncom.2013.00174/abstract
7. Glimcher PW (2011) Understanding dopamine and reinforcement learning: the dopamine reward prediction error hypothesis. Proc Natl Acad Sci 108:15647–15654
8. Wickens J, Kotter R (1995) Cellular models of reinforcement. In: Houks JC, Davis JL, Beiser DG (eds) Models of information processing in the basal ganglia. MIT Press, Cambridge
9. Monchi O, Taylor JG, Dagher A (2000) A neural model of working memory processes in normal subjects, Parkinson's disease and schizophrenia for fMRI design and predictions. Neural Netw 13:953–973
10. Heaton RK, Chelune GJ, Talley JL, Kay GG, Curtiss G (1993) Wisconsin card sorting test manual. Psychological Assessment Resources, Inc., Odessa
11. Frank MJ (2005) Dynamic dopamine modulation in the basal ganglia: a neurocomputational account of cognitive deficits in medicated and nonmedicated Parkinsonism. J Cogn Neurosci 17:51–72
12. Ashby FG, Alfonso-Reese LA, Turken AU, Waldron EM (1998) A neuropsychological theory of multiple systems in category learning. Psychol Rev 105(3):442–481
13. Ashby FG, Valentin VV (2017) Multiple systems of perceptual category learning: theory and cognitive tests. In: Cohen H, Lefebvre C (eds) Handbook of categorization in cognitive science, 2nd edn. Elsevier, Oxford, pp 157–188
14. Hélie S, Paul EJ, Ashby FG (2012) A neurocomputational account of cognitive deficits in Parkinson's disease. Neuropsychologia 50(9):2290–2302
15. Hélie S, Paul EJ, Ashby FG (2012) Simulating the effects of dopamine imbalance on cognition: from positive affect to Parkinson's disease. Neural Netw 32:74–85

16. Pasti L, Volterra A, Pozzan T, Carmignoto G (1997) Intracellular calcium oscillations in astrocytes: a highly plastic, bidirectional form of communication between neurons and astrocytes in situ. J Neurosci 17(20):7817–7830
17. Araque A, Carmignoto G, Haydon PG (2001) Dynamic signaling between astrocytes and neurons. Ann Rev Physiol 63(1):795–813
18. Perea G, Navarrete M, Araque A (2009) Tripartite synapses: astrocytes process and control synaptic information. Trends Neurosci 32(8):421–431
19. Rusakov DA, Zheng K, Henneberger C (2011) Astrocytes as regulators of synaptic function a quest for the ca2+ master key. The Neuroscientist 17(5):513–523
20. Haydon PG (2001) Glia: listening and talking to the synapse. Nat Rev Neurosci 2(3):185–193
21. Liddelow SA, Guttenplan KA, Clarke LE, Bennett FC, Bohlen CJ, Schirmer L, Bennett ML, Münch AE, Chung WS, Peterson TC et al (2017) Neurotoxic reactive astrocytes are induced by activated microglia. Nature 541(7638):481–487
22. Maragakis NJ, Rothstein JD (2006) Mechanisms of disease: astrocytes in neurodegenerative disease. Nat Clin Pract Neurol 2(12):679–689
23. Jellinger KA (2010) Basic mechanisms of neurodegeneration: a critical update. J Cell Mol Med 14(3):457–487
24. Forman MS, Lal D, Zhang B, Dabir DV, Swanson E, Lee VMY, Trojanowski JQ (2005) Transgenic mouse model of tau pathology in astrocytes leading to nervous system degeneration. J Neurosci 25(14):3539–3550
25. Lobb C (2014) Abnormal bursting as a pathophysiological mechanism in Parkinson's disease. Basal Ganglia 3(4):187–195
26. Izhikevich EM (2007) Dynamical systems in neuroscience. Cambridge, MA, MIT Press
27. Hodgkin AL (1951) The ionic basis of electrical activity in nerve and muscle. Biol Rev 26(4):339–409
28. Porto-Pazos AB, Veiguela N, Mesejo P, Navarrete M, Alvarellos A, Ibáñez O, Pazos A, Araque A (2011) Artificial astrocytes improve neural network performance. PloS One 6(4):e19109
29. Alvarellos-González A, Pazos A, Porto-Pazos AB (2012) Computational models of neuron-astrocyte interactions lead to improved efficacy in the performance of neural networks. Comput Math Methods Med 2012:10
30. Sajedinia Z (2014) Artificial astrocyte networks, as components in artificial neural networks. In: Unconventional computation and natural computation. Springer, pp 316–326
31. Nadkarni S, Jung P (2004) Dressed neurons: modeling neural–glial interactions. Phys Biol 1(1):35
32. Valenza G, Tedesco L, Lanatà A, De Rossi D, Scilingo EP (2013) Novel spiking neuron-astrocyte networks based on nonlinear transistor-like models of tripartite synapses. In: 2013 35th annual international conference of the IEEE engineering in medicine and biology society (EMBC). IEEE, pp 6559–6562
33. Haghiri S, Ahmadi A, Nouri M, Heidarpur M (2014) An investigation on neuroglial interaction effect on Izhikevich neuron behaviour. In: 2014 22nd Iranian conference on electrical engineering (ICEE). IEEE, pp 88–92
34. Sajedinia Z, Hélie S. (2018) A new computational model for astrocytes and their role in biologically realistic neural networks. Comput Intell Neurosci 2018
35. Pannasch U, Vargová L, Reingruber J, Ezan P, Holcman D, Giaume C, Syková E, Rouach N (2011) Astroglial networks scale synaptic activity and plasticity. Proc Natl Acad Sci 108(20):8467–8472
36. Hélie S, Fleischer P (2016) Simulating the effect of reinforcement learning on neuronal synchrony and periodicity in the striatum. Front Comput Neurosci 10:40
37. Ponzi A, Wickens J (2010) Sequentially switching cell assemblies in random inhibitory networks of spiking neurons in the striatum. J Neurosci 30:5894–5911

Chapter 11
Attentional Deficits in Alzheimer's Disease: Investigating the Role of Acetylcholine with Computational Modelling

Eirini Mavritsaki, Howard Bowman, and Li Su

Abstract Alzheimer's disease is a neurodegenerative condition that affects the brain's cognitive processes as well as many other functions for daily life. It is the commonest cause for dementia in older people and can take several years or decades from the time its pathology starts to the time the full clinical symptoms are developed. One of the cognitive processes affected in Alzheimer's disease is attention. Depletion in attentional processes is linked to acetylcholine function, and attention deficit underlies many cognitive dysfunctions in Alzheimer's disease. In this work, we are employing computational modelling to provide a neural bio-mechanistic account linking acetylcholine depletion and decreased attentional performance. Although previous research has modelled the decrease of acetylcholine, how neurotransmitter depletion is associated with behavioural impairments in Alzheimer's disease remains unclear. We employed a spiking Search over Time and Space (sSoTS) model to simulate attentional function and describe the reduction of acetylcholine by changes applied to *gNMDA* and *gAMPA* conductance. Our model simulation results showed that changes in acetylcholine function were able to produce a notable reduction in attentional performance similar to what is seen in patients with Alzheimer's disease. This work provided an architectural and

E. Mavritsaki (✉)
Department of Psychology, Birmingham City University, Birmingham, UK

School of Psychology, University of Birmingham, Birmingham, UK
e-mail: Eirini.mavritsaki@bcu.ac.uk

H. Bowman
School of Psychology, University of Birmingham, Birmingham, UK

School of Computing, University of Kent, Canterbury, UK
e-mail: H.Bowman@kent.ac.uk

L. Su
Department of Psychiatry, University of Cambridge, Cambridge, UK

Sino-Britain Centre for Cognition and Ageing Research, Southwest University, Chongqing, China
e-mail: ls514@cam.ac.uk

© Springer Nature Switzerland AG 2019   113
V. Cutsuridis (ed.), *Multiscale Models of Brain Disorders*, Springer Series in
Cognitive and Neural Systems 13, https://doi.org/10.1007/978-3-030-18830-6_11

methodological framework under which neurobiological mechanisms and failures of the system can directly explain symptomology, such as attention dysfunctions in Alzheimer's disease. This framework enables future studies and novel clinical trials targeting acetylcholine pathways in treating Alzheimer's disease and related conditions.

Keywords Alzheimer's · Spiking Search over Time and Space model · Computational modelling · Attention · Acetylcholine

Attention is a very important cognitive process that is employed for many actions in our everyday life (e.g. watching television, reading a paper, washing our face, eating and so on). It is therefore essential to investigate further the underlying mechanisms in neurodegenerative conditions, like Alzheimer's disease, in which our attentional abilities are reduced [15, 16, 21, 41, 42, 50, 53]. Alzheimer's disease is a condition that can take several years if not decades from the time it starts to the time the full symptoms are shown [51]. In those years of disease progression, there are a number of pathological processes that are taking place; however, one of the starting points of the pathology is believed to be the aggregation of β-amyloids into plaques [19, 51]. Irrespective of the amount of research that has taken place, many questions remain on how the disease unfolds and how to identify individual's position in the disease's trajectory [19, 45].

It is therefore essential to fully understand the underlying processing and progression mechanisms of the pathology. Understanding the progression of the pathology will allow for early diagnosis and improve medications and managements providing better and longer life for patients. In fact to achieve this, it has been argued that cognitive tasks related to cognitive impairments are reliable measures to predict progression of the disease [4]. As previously discussed, one of the cognitive functions that is affected in Alzheimer's disease is attention [15, 16, 21, 41, 42, 50, 53]. Patients with Alzheimer's disease have been shown to have reduced visual processing and attentional orienting [26] that affect their attentional processing and visual search [9, 26].

To investigate further attentional processing, researchers have been using the single feature and conjunction search tasks [37, 42, 54, 56]. In both experiments, subjects are asked to identify a target amongst distractors. In single feature search, the target shares one common feature with distractors, for example, a target blue *H* amongst blue *A* distractors. In the conjunction task, the target shares two different features with the distractors, for example, a target blue *H* amongst distractors that are blue *As* and green *Hs* [54, 56]. Patients with Alzheimer's disease show reduced activation in visual, dorsal attention and ventral attention networks [26] leading to expected deficits in attentional processes [15, 16, 21, 41, 42, 50, 53]. Interestingly, the patients show a significant decrease in performance in the difficult conjunction task but no significant changes in the easy single feature search [10, 16, 21, 40, 49, 50]. Researchers suggest that this might be due to an impairment in the binding and grouping processes [21], problems with inhibitory processes

[24, 38] or acetylcholine changes that are linked to Alzheimer's disease [5, 11, 25, 34, 48, 55]. Acetylcholine is directly linked to attentional processes [12, 22, 25, 46], through its cholinergic projections from the basal forebrain [22]. Pyramidal neurons modulated by acetylcholine express nicotinic and muscarinic receptors [22]. Acetylcholine modulation from the nucleus basalis magnocellularis is linked to changes in binding in visual search tasks [6], and changes in feature binding affect the difficult rather than easy visual search [52]. Furthermore, acetylcholine modulation affects attentional processes through gamma synchrony [12], which is linked with changes observed in oscillatory activity in Alzheimer's disease. In fact, Deco and Thiele [12] identified that there is an optimum ratio between NMDA and AMPA conductance that is linked to acetylcholine, gamma oscillations and optimum attention.

All cognitive processes rely on rapidly occurring coordinated action among distributed neural assemblies. Thus, one approach to understand and identify when and how these processes go wrong is to measure coordinated neural activity in brain networks. Neural oscillations, as observed with EEG or MEG (M/EEG), are such a measure. Brain oscillations represent regular fluctuations in electrical potentials/magnetic fields and are generated by tens of thousands of neurons. Oscillations occur at different frequencies, ranging from very slow (0.2 Hz) to very fast (300 Hz), where the frequency of an oscillation is thought to inversely reflect the size of the network generating the oscillation. Brain oscillations thus capture the fundamental characteristics of the structural wiring of the brain and allow neural communication during different cognitive states (rest, perception, memory, etc.) to be explored at the temporal resolution at which neurons operate. For this reason, brain oscillations are a promising candidate for charting neurological and psychiatric disorders [8, 17, 20, 43, 59].

There is an expanding literature on MEG/EEG (M/EEG) correlates of a range of disorders, such as epilepsy, memory impairments, OCD and ADHD [20]. Of particular relevance, there are now a number of findings on oscillatory correlates of mild cognitive impairment (MCI) (e.g. Gomez et al. [18]), Alzheimer's (e.g. Stam et al. [47]) and other dementias (e.g. Hughes and Rowe [23]), including in large multisite studies [27].

Additionally, in the resting-state brain (in which stationarity of oscillatory features can be assumed), a number of interesting findings have been reported. For example, one hallmark of dementia, particularly Alzheimer's, is a general slowing of the resting-state frequency spectrum [36]. Additionally, Poil et al. [39] found a wide spectrum of EEG resting-state measures that predicted progression to Alzheimer's, with a broader beta power peak most predictive [39].

The oscillatory correlates of MCI are perhaps particularly important for attention as well, since it is frequently a precursor diagnosis to Alzheimer's. As a result, its detection is a target for work on biomarkers of the very early stages of Alzheimer's. With this goal in mind, an oscillatory EEG pattern that distinguishes mild cognitive impairment (MCI) patients who either progress within 3 years (convertors) or do not (stable) to Alzheimer's disease (AD) has been identified [33]. Figure 11.1 shows time-frequency spectra for brain activity arising from lexical processing.

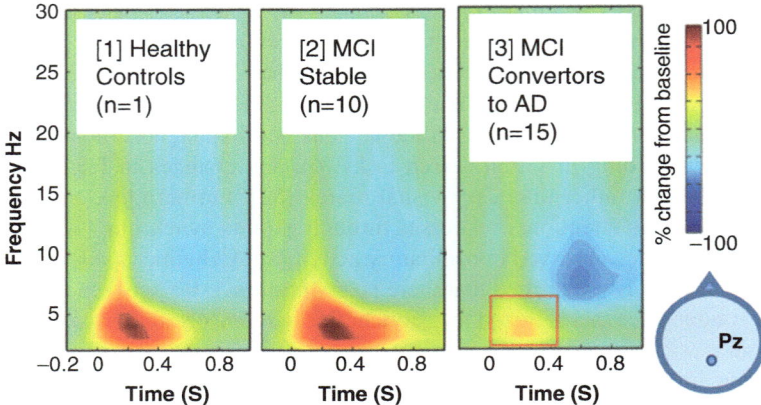

Fig. 11.1 An attenuation of theta activity associated with lexical processing in a group of MCI patients who would go on to convert to Alzheimer's disease [33]. The time-frequency spectra are locked to word onset at the midline-parietal electrode Pz. There is a theta increase 0–0.5 s after the onset of the word, for the healthy controls (panel [1]) and for the MCI patients who did not convert to Alzheimer's disease (i.e. MCI Stable, panel [2]). This theta increase was significantly attenuated in MCI patients who would later convert to Alzheimer's disease (panel [3]). For all groups, word presentation is followed by an alpha suppression effect (0.5–1 s)

Importantly, healthy controls (panel [1]) and non-progressors (i.e. MCI-stable, panel [2]) show a clear increase in theta power 0–0.5 s after word onset (warm colours), while MCI-convertors (panel [3]) show a reduced increase (convertors<stable: $p < 0.046$; convertors<control: $p < 0.004$). Such an oscillatory change for lexical processing is consistent with language deficits in AD [14].

A key question that follows is the effect of neuromodulators on these EEG patterns. As previously discussed, there is considerable evidence that there are a range of changes to neuromodulators associated with the development of dementias. In this respect, changes in acetylcholine are of particular interest. There are, though, few studies that explicitly consider differences in human EEG features between groups with and without Alzheimer's when Acetylcholine is manipulated.

One of the few studies that explicitly targeted this question is by Yener and collaborators [57]. They compared healthy controls with Alzheimer's patients on and off AchEI, a compound that increases the level and duration of action of acetylcholine. The authors found a reduction in phase locking (to stimulus presentation) of theta oscillations at a frontal electrode for the Alzheimer's group that were not on AchEl, i.e. who, it is assumed, had depleted acetylcholine. It is notable that theta was the relevant oscillation both in the Yener et al. [57] and the Mazaheri et al. [33] studies. This said, the former focussed on phase coherence across replications, while the latter focussed on power changes, and electrode sites were different. Nonetheless, the link between depleted acetylcholine in Alzheimer's and a noisier stimulus locked theta oscillation is definitely worth further exploration.

An attractive way to proceed in this line of research is computational modelling, which allows data collected from all different methodologies to be combined, to test outcomes and to provide predictions for further testing [2, 12, 26, 29, 32]. It is very important therefore to use computational modelling to help us interpret the findings so far and to progress further. Moreover, developed computational models could then be used to predict the progression of the disease on an individual basis and predict the efficacy of different drug targets in Alzheimer's disease if the model captures the mechanism of these drug compounds in the patients. If the model parameters can be determined at individual rather than group level, such model paves the way for personalized treatments.

Accordingly, researchers have started modelling Alzheimer's disease [1, 2, 58], but they have focussed on low-level properties of the system, rather than linking neurophysiological damage with behaviour. In contrast, in the work presented here, we use a computational model that can allow these two levels to be linked and allow us not only to understand Alzheimer's at the neuronal level but also how changes observed in Alzheimer's disease are linked with behavioural changes. The selected behavioural study is visual search. This is because, there is a good deal of work in the area that identifies depletion in attentional processes in Alzheimer's disease and links it to acetylcholine function, and attention deficit underlies many cognitive dysfunctions in Alzheimer's disease [12, 15, 16, 21, 41, 42, 50, 53]. Furthermore, this work is based on the binding spiking Search over Time and Space (bsSoTS) [29] that has been extensively used to simulate visual attention processes in healthy adults [30–32] and to investigate further the attentional processes in conditions where such processes are depleted [29, 31, 32]. The bsSoTS is also the appropriate model to use because the parameters of the model have been set to generate neural activity resembling that of the human brain in the content of realistic noise component. To simulate the acetylcholine depletion, the work of Deco and Thiele [12] is followed where the attentional behaviour changes are investigated by changes in acetylcholine levels through the AMPA and NMDA currents.

11.1 Methods

The methodology presented in this work is based on the bsSoTS model that uses integrate-and-fire neurons to simulate the traditionally used visual search experiment [29, 32]. Neuronal properties are described in Mavritsaki and Humphreys [28] based on the integrate-and-fire neurons of Brunel and Wang [7]. Input to the cell is based on a fast excitatory AMPA current, a slow excitatory NMDA current, an inhibitory GABA current and a frequency adaptation based on the calcium-sensitive potassium current I_{AHP}. The model is ideal for this level of simulations as it has successfully simulated the easy and difficult visual search experiment and incorporates top-down and bottom-up processes [29, 32], as well as allowing us to investigate neuronal changes dependent on the AMPA, GABA and NMDA

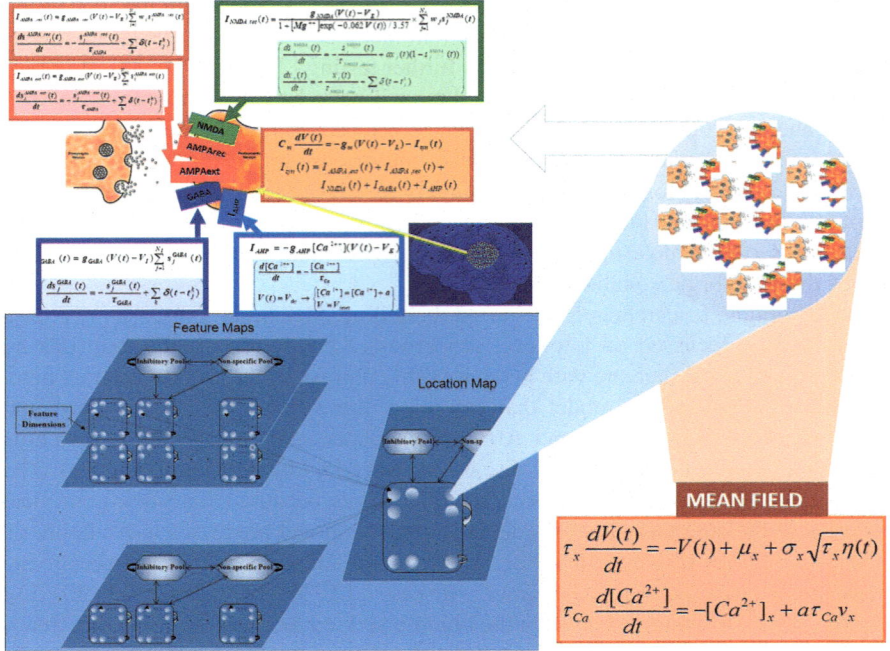

Fig. 11.2 bsSoTS organisation for a range of feature maps. For the results presented in this work, we are using two feature maps for shape and colour. The model is overall separated into three layers, two layers for encoding the feature characteristics (the two feature maps) and a third layer in which all information is combined, the location map [31]. To constrain the model, we move through two different levels: level one is the mean field, where a group of neurons is simulated using a transfer function as shown in Fig. 11.2, and the spiking level, where each neuron is simulated using the equations shown on the top left corner of the figure. For more details on the model, please see Mavritsaki and colleagues work [29, 30, 32]

currents [28]. The organisation of the model is based on previous work by Deco and Zihl [13] and follows Feature Integration Theory [52]. The model simulates visual search experiments as described above. The general organisation of the model and the spiking and mean-field neuronal level is presented in Fig. 11.2; to model the simulated experiment, the model is divided into three layers: two feature layers whose activation is bound into the location map/saliency map [31]. Each feature dimension layer is separated into two feature maps: the shape feature layer is separated into H and A and the colour feature layer is separated into colour blue and colour green, as in previous work [28, 29, 32].

Following the work by Deco and Thiele on the effects of acetylcholine on attention [12], we investigated the effects of changes in g_{NMDA} and g_{AMPA} on attentional processes in an effort to simulate the changes in visual search, assuming only changes in acetylcholine. We changed the AMPA and NMDA conductance from performance observed in Alzheimer's −16 to 16%. Figure 11.3 (gAMPA/gNMDA

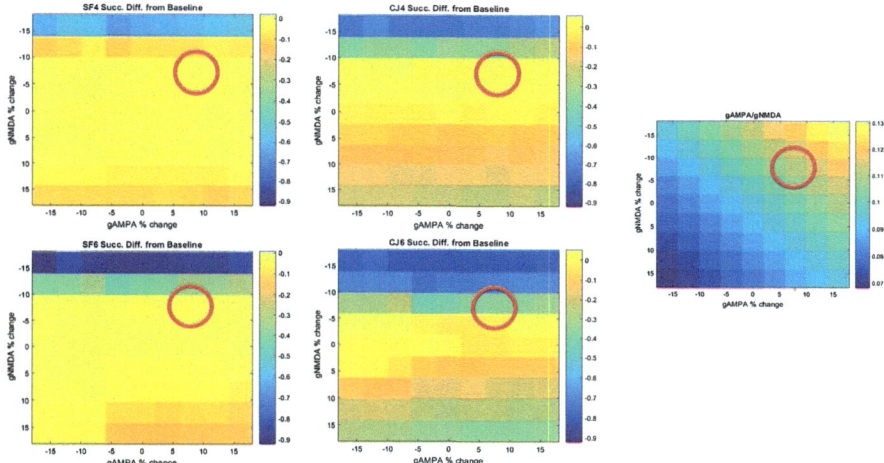

Fig. 11.3 On the left we show the success rate differences from baseline success rate for the single feature 4 (SF4) items, single feature 6 (SF6) items, conjunction 4 (CJ4) items and conjunction 6 (CJ6) items. The baseline values that are used for the calculations are 95% for SF4 items, 96% for SF6 items, 81% for CJ4 items and 80% for CJ6 items. The circle parameter group is the parameter group that was identified as optimum to simulate Alzheimer's visual search behaviour. On the left we show AMPA/NMDA conductance ratio changes for the parameter space used. The circle marks the set of parameters that was identified as the optimal to simulate Alzheimer's visual search behaviour

graph) presents the effect of the changed values to the ratio of the NMDA/AMPA conductance [12]. Within this range of parameters, we identified the parameters that allow for a different decline in visual search performance to be observed for simulated Alzheimer's disease between difficult and easy conditions. From the range of parameters investigated, the parameter settings that allowed for a greater decline in performance for the difficult relative to single feature search task were selected. The single feature and conjunction visual search experiments were simulated for this selected set of parameters.

The Poisson noise presented to the model allows us to simulate human performance by running the model for 300 trials for each display size (4 and 6) in single feature and conjunction conditions. This analysis follows the same analysis that was previously performed [28, 29, 32]. The reaction times (RTs) and success rates obtained were then analysed using a mixed ANOVA design.

11.2 Results

The RTs and success rates for all the gNMDA and gAMPA parameter changes presented in Fig. 11.3 are calculated for each parameter set and presented in Figs. 11.3 and 11.4. The changes for RTs and success rates for single feature

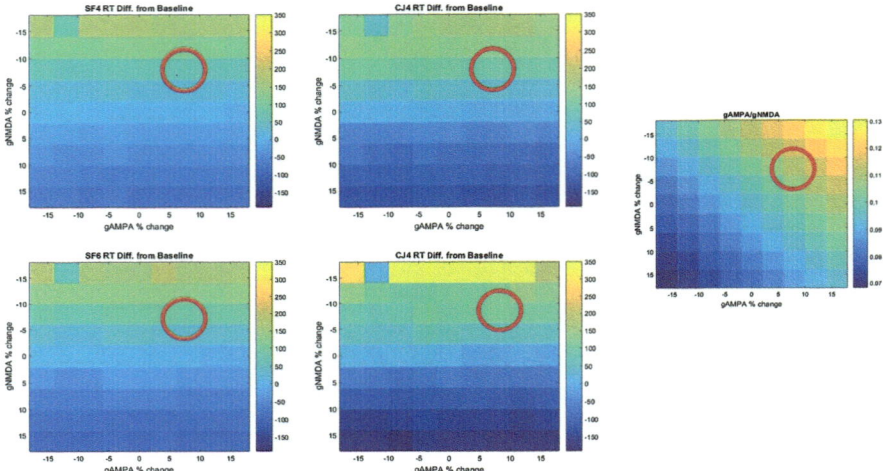

Fig. 11.4 On the left we show reaction time difference from baseline reaction time for the single feature 4 (SF4) items, single feature 6 (SF6) items, conjunction 4 (CJ4) items and conjunctions 6 (CJ6) items. The baseline values that are used for the calculations are 254.5 ms for SF4 items, 299.9 ms for SF6 items, 337.2 ms for CJ4 items and 432.9 ms for CJ6 items. The circle marks the parameter group that was identified as optimum to simulate Alzheimer's visual search behaviour. On the left we show again the AMPA/NMDA conductance ratio changes for the parameter space used.

were smaller than the changes for conjunction. The circled parameter set identified in Figs. 11.3 and 11.4 is the one selected to simulate the Alzheimer's condition. Figure 11.5 illustrates the average RTs, and Fig. 11.6 illustrates the average success rates.

Simulated participants' RTs and success rates across all conditions were entered into a $2 \times 2 \times 2$ mixed-design ANOVA with the within-participants factor of condition (single feature/conjunction search) and display size (4/6 items) and the between-participants factor of group (simulated control/simulated AD groups). This gave us the main effects and interactions presented in Table 11.1. A significant interaction of condition x display size x group was observed for percentage success rate, but not for reaction times. For reaction time, we observed significant main effects of display size, $F(1,28) = 181$, $p < 0.001$ and $\eta_p^2 = 0.866$; condition, $F(1,28) = 243.4$, $p < 0.001$ and $\eta_p^2 = 0.897$; group $F(1,28) = 276.1$, $p < 0.001$ and $\eta_p^2 = 0.908$; and an interaction of condition with display size, $F(1,28) = 13.9$, $p = 0.001$ and $\eta_p^2 = 0.332$. We did not observe significant interactions between display size and group, $F(1,28) = 1.7$, $p = 0.191$ and $\eta_p^2 = 0.06$, or condition and group, $F(1,28) = 1.4$, $p = 0.236$ and $\eta_p^2 = 0.05$. For success rate, we observed significant main effects of display size, $F(1,28) = 67.2$, $p < 0.001$ and $\eta_p^2 = 0.706$; condition, $F(1,28) = 222.2$, $p < 0.001$ and $\eta_p^2 = 0.888$; and group, $F(1,28) = 34.2$,

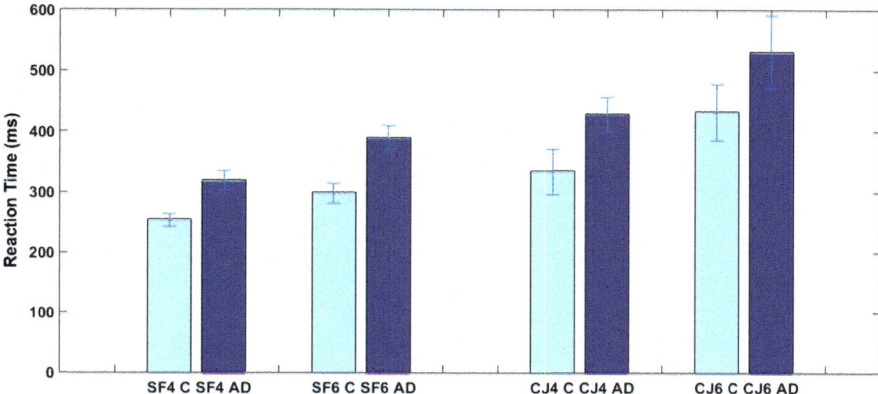

Fig. 11.5 Bar plots for reaction times for simulated Alzheimer's and baseline (simulated controls). SF4 C shows the reaction time for single feature 4 items for controls (healthy participants); SF4 AD shows the reaction time for single feature 6 items for simulated Alzheimer's patients. SF6 C shows the reaction time for single feature 6 items for controls; SF6 AD shows the reaction time for single feature 6 items for simulated Alzheimer's patients. The same applies for conjunction, where CJ4 is conjunction for 4 items and CJ6 is conjunction for 6 items

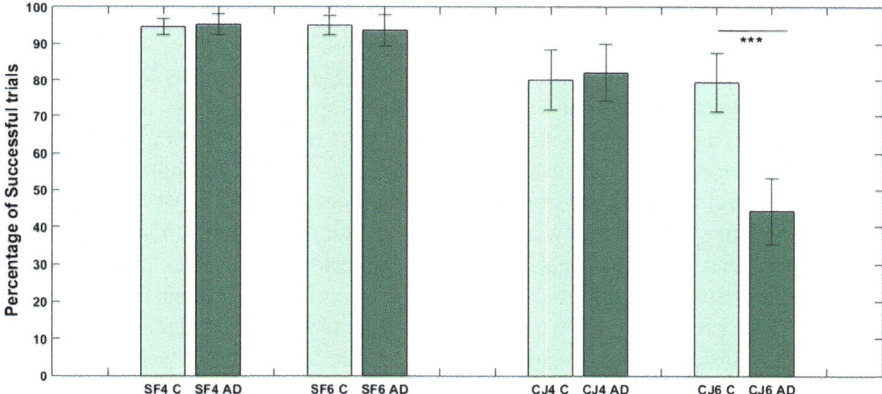

Fig. 11.6 Bar plots for success rate for simulated Alzheimer's and Baseline (simulated controls). SF4 C shows the success rate for single feature 4 items for controls (healthy participants); SF4 AD shows the success rate for single feature 6 items for simulated Alzheimer's patients. SF6 C shows the success rate for single feature 6 items for controls; SF6 AD shows the success rate for single feature 6 items for simulated Alzheimer's patients. The same applies for conjunction where CJ4 is conjunction for 4 items and CJ6 is conjunction for 6 items

$p < 0.001$ and $\eta_p^2 = 0.551$. We also observed significant interactions between condition and display size, $F(1,28) = 59.2$, $p < 0.001$ and $\eta_p^2 = 0.682$; condition and group, $F(1,28) = 26.4$, $p < 0.001$ and $\eta_p^2 = 0.486$; and display size and group, $F(1,28) = 65$, $p < 0.001$ and $\eta_p^2 = 0.699$.

Table 11.1 ANOVAs of percentage success and reaction times, reporting the main effects of and interactions between condition (single feature/conjunction search), display size (4/6 items) and group (baseline/AD group)

Measure	ANOVA term	F (*df*)	*p*	η_p^2
Reaction time	Display size	181 (1,28)	<0.001	0.866
% Success	Condition	243.4 (1,28)	<0.001	0.897
	Group	276.1 (1,28)	<0.001	0.908
	Condition × display size	13.9 (1,28)	0.001	0.332
	Condition × group	1.4 (1,28)	0.236	0.05
	Display size × group	1.7 (1,28)	0.191	0.06
	Condition × display size × group	0.8 (1,28)	0.36	0.03
	Display size	67.2	<0.001	0.706
	Condition	222.2	<0.001	0.888
	Group	34.2	<0.001	0.551
	Condition × display size	59.9	<0.001	0.682
	Condition × group	26.4	<0.001	0.486
	Display size × group	65	<0.001	0.699
	Condition × display size × group	53.6	<0.001	0.657

11.3 Discussion

This study fits broadly into the category of Computational Psychiatry [35, 58]. Although majority of Computational Psychiatry approaches aim to decompose behaviour into its constituents before explaining its neural underpinnings, such as mapping decision-making into values and prediction errors. Our approach aimed at providing an architectural framework under which neurobiological dysfunctions can directly explain symptomology in Alzheimer's. The results from this work clearly demonstrate that computational modelling that bridges low-level characteristics with whole system behaviour can be used to simulate attentional processes, as has also been previously shown [32, 44], and to make that bridge in simulating Alzheimer's disease changes in attentional processing. There were significant changes observed between simulated controls and simulated Alzheimer's disease patients for success rates and reaction times, as has been previously shown [10, 16, 21, 40, 49, 50] and significant interactions for condition, display size and group. This clearly demonstrates that the model was able to capture some of the changes in Alzheimer's disease attentional processing by changing the gNMDA and gAMPA parameters, simulating reduction in acetylcholine observed in Alzheimer's disease.

Although the results showed significant overall interactions for success rates, the overall interaction for reaction time was not significant. This can be attributed to the fact that the healthy simulated behaviour is at ceiling level; therefore any changes due to the acetylcholine reduction might not be easily detected. Another reason might be that although acetylcholine reduction is important in Alzheimer's disease there are other processes that might also be affected, like changes in binding, grouping or inhibitory processes [21, 24, 38]. Changes in these processes are not taken into consideration in the present work, as this work aimed simply to investigate if acetylcholine changes could be simulated using the chosen approach. Although these processes were not investigated in the current work, the model incorporates all the above processes [29, 32] and therefore can allow us to further investigate them in the following steps of our work. Furthermore, additional work is required to investigate where acetylcholine changes can be related to the oscillatory behaviour in Alzheimer's disease demonstrated by Mazaheri et al. [33], thereby helping to shed light on further understanding of disease progression. The presented work therefore demonstrated that the proposed and similar modelling approaches can be used to simulate Alzheimer's disease. The following steps are to use the model to combine the changes in attentional processes which are found in Alzheimer's disease in one model and use the model to shed light on this condition. Furthermore, if this is combined with oscillatory behaviour study [33] by using previously develop approaches for extracting MEG activity from similar models [3], it may be able to provide the first step for delivering personalized treatment in Alzheimer's disease.

References

1. Adeli H, Ghosh-Dastidar S, Dadmehr N (2005) Alzheimer's disease: models of computation and analysis of EEGs. Clin EEG Neurosci 36(3):131–140. Retrieved from http://www.scopus.com/inward/record.url?eid=2-s2.0-23844473285&partnerID=40&md5=c54e220ac7d6d50a5afe85f0d2357095
2. Alexiou A, Mantzavinos VD, Greig NH (2017) A Bayesian model for the prediction and early diagnosis of Alzheimer' s disease. Front Aging Neurosci 9:1–14. https://doi.org/10.3389/fnagi.2017.00077
3. Barbieri F, Mazzoni A, Logothetis NK, Panzeri S, Brunel N (2014) Stimulus dependence of local field potential spectra: experiment versus theory. J Neurosci 34(44):14589–14605. https://doi.org/10.1523/JNEUROSCI.5365-13.2014
4. Belleville S, Fouquet C, Hudon C, Zomahoun HTV, Croteau J (2017) Neuropsychological measures that predict progression from mild cognitive impairment to Alzheimer's type dementia in older adults: a systematic review and meta-analysis. Neuropsychol Rev 27(4):328–353. https://doi.org/10.1007/s11065-017-9361-5
5. Bertrand D, Jr AVT (2018) The wonderland of neuronal nicotinic acetylcholine receptors. Biochem Pharmacol 151:214–225. https://doi.org/10.1016/j.bcp.2017.12.008
6. Botly LCP, De Rosa E (2012) Impaired visual search in rats reveals cholinergic contributions to feature binding in visuospatial attention. Cereb Cortex 22(10):2441–2453. https://doi.org/10.1093/cercor/bhr331
7. Brunel N, Wang XJ (2001) Effects of neuromodulation in a cortical network model of object working memory dominated by recurrent inhibition. J Comput Neurosci 11(1):63–85. https://doi.org/10.1023/A:1011204814320
8. Colom LV, García-Hernández A, Castañeda MT, Perez-Cordova MG, Garrido-Sanabria ER (2006) Septo-hippocampal networks in chronically epileptic rats: potential antiepileptic effects of theta rhythm generation. J Neurophysiol 95(6):3645–3653. https://doi.org/10.1152/jn.00040.2006
9. Corbetta M, Patel G, Shulman GL (2008) The reorienting system of the human brain: from environment to theory of mind. Neuron. https://doi.org/10.1016/j.neuron.2008.04.017

10. Cormack F, Gray A, Ballard C, Tovée MJ (2004) A failure of "pop-out" in visual search tasks in dementia with Lewy bodies as compared to Alzheimer's and Parkinson's disease. Int J Geriatr Psychiatry 19(8):763–772. https://doi.org/10.1002/gps.1159

11. Daiello LA, Pharm D, Ott BR, Festa EK, Friedman M, Miller LA, Heindel WC (2010) Effects of cholinesterase inhibitors on visual attention in drivers with Alzheimer's disease. J Clin Psychopharmacol 30(3):245–251. https://doi.org/10.1097/JCP.0b013e3181da5406.Effects

12. Deco G, Thiele A (2009) Attention – oscillations and neuropharmacology. Eur J Neurosci 30(3):347–354. https://doi.org/10.1111/j.1460-9568.2009.06833.x

13. Deco G, Zihl J (2001) Top-down selective visual attention: a neurodynamical approach. Vis Cogn 8(1):119–140

14. Ferris SH, Farlow M (2013) Language impairment in Alzheimer's disease and benefits of acetylcholinesterase inhibitors. Clin Interv Aging 8:1007–1014. https://doi.org/10.2147/CIA.S39959

15. Festa EK, Heindel WC, Ott BR (2010) Dual-task conditions modulate the efficiency of selective attention mechanisms in Alzheimer's disease. Neuropsychologia. https://doi.org/10.1016/j.neuropsychologia.2010.07.003

16. Foster JK, Behrmann M, Stuss DT (1999) Visual attention deficits in Alzheimer's disease: simple versus conjoined feature search. Neuropsychology 13(2):223–245

17. Gallego-Jutglà E, Solé-Casals J, Vialatte F-B, Dauwels J, Cichocki A (2014) A Theta-band EEG based index for early diagnosis of Alzheimer's disease. J Alzheimers Dis 43(4):1175–1184. https://doi.org/10.3233/JAD-140468

18. Gomez C, Stam CJ, Hornero R, Fernandez A, Maestu F (2009) Disturbed beta band functional connectivity in patients with mild cognitive impairment: an MEG study. IEEE Trans Biomed Eng 56(6):1683–1690. https://doi.org/10.1109/TBME.2009.2018454

19. Gordon BA, Blazey TM, Su Y, Hari-raj A, Dincer A, Flores S et al (2018) Spatial patterns of neuroimaging biomarker change in individuals from families with autosomal dominant Alzheimer's disease: a longitudinal study. Lancet Neurol:241–250. https://doi.org/10.1016/S1474-4422(18)30028-0

20. Güntekin B (2008) A review of brain oscillations in cognitive disorders and the role of neurotransmitters. Brain Res 1235:172–193. https://doi.org/10.1016/J.BRAINRES.2008.06.103

21. Hao J, Li K, Li K, Zhang D, Wang W, Yang Y et al (2005) Visual attention deficits in Alzheimer's disease: an fMRI study. Neurosci Lett. https://doi.org/10.1016/j.neulet.2005.05.028

22. Hedrick T, Waters J (2018) Acetylcholine excites neocortical pyramidal neurons via nicotinic receptors. J Neurophysiol 113:2195–2209. https://doi.org/10.1152/jn.00716.2014

23. Hughes LE, Rowe JB (2013) The impact of neurodegeneration on network connectivity: a study of change detection in frontotemporal dementia. J Cogn Neurosci 25(5):802–813. https://doi.org/10.1162/jocn_a_00356

24. Levinoff EJ, Li KZH, Murtha S, Chertkow H (2004) Selective attention impairments in Alzheimer's disease: evidence for dissociable components. Neuropsychology 18(3):580–588. https://doi.org/10.1037/0894-4105.18.3.580

25. Levy JA, Parasuraman R, Greenwood PM, Dukoff R, Sunderland T (2000) Acetylcholine affects the spatial scale of attention: evidence from Alzheimer's disease. Neuropsychology 14(2):288–298. https://doi.org/10.1037//0894-4105.14.2.288

26. Li H-J, Hou X-H, Liu H-H, Yue C-L, He Y, Zuo X-N (2015) Toward systems neuroscience in mild cognitive impairment and Alzheimer's disease: a meta-analysis of 75 fMRI studies. Hum Brain Mapp 36(3):1217–1232. https://doi.org/10.1002/hbm.22689

27. Maestú F, Peña J-M, Garcés P, González S, Bajo R, Bagic A et al (2015) A multicenter study of the early detection of synaptic dysfunction in mild cognitive impairment using magnetoencephalography-derived functional connectivity. NeuroImage: Clin 9:103–109. https://doi.org/10.1016/J.NICL.2015.07.011

28. Mavritsaki E, Humphreys GW (2013) Different functional roles of dopamine and acetylcholine in visual selection: simulations of visual search in Alzheimer's and Parkinson's diseases. 2J Vision 13:523. https://doi.org/10.1167/13.9.523

29. Mavritsaki E, Humphreys G (2016) Temporal binding and segmentation in visual search: a computational neuroscience analysis. J Cogn Neurosci 28(10):1553–1567. https://doi.org/10.1162/jocn_a_00984

30. Mavritsaki E, Heinke D, Humphreys GW, Deco G (2006) A computational model of visual marking using an inter-connected network of spiking neurons: the spiking search over time & space model (sSoTS). Journal of Physiology, Paris 100:110–124. https://doi.org/10.1016/j.jphysparis.2006.09.003

31. Mavritsaki E, Allen HA, Humphreys GW (2010) Decomposing the neural mechanisms of visual search through model-based analysis of fMRI: top-down excitation, active ignoring and the use of saliency by the right TPJ. NeuroImage 52(3):934–946. https://doi.org/10.1016/j.neuroimage.2010.03.044

32. Mavritsaki E, Heinke D, Allen H, Deco G, Humphreys GW (2011) Bridging the gap between physiology and behavior: evidence from the sSoTS model of human visual attention. Psychol Rev 118(1):3–41. https://doi.org/10.1037/a0021868

33. Mazaheri A, Segaert K, Olichney J, Yang J-C, Niu Y-Q, Shapiro K, Bowman H (2018) EEG oscillations during word processing predict MCI conversion to Alzheimer's disease. NeuroImage: Clin 17:188–197. https://doi.org/10.1016/J.NICL.2017.10.009

34. Mesulam M (2004) The cholinergic lesion of Alzheimer's disease: pivotal factor or side show? Learn Mem 11(1):43–49. https://doi.org/10.1101/lm.69204

35. Montague PR, Dolan RJ, Friston KJ, Dayan P (2012) Computational psychiatry. Trends Cogn Sci 16(1):72–80. https://doi.org/10.1016/J.TICS.2011.11.018

36. Montez, T., Poil, S., . . . B J.-P. of the, & 2009, undefined (n.d.) Altered temporal correlations in parietal alpha and prefrontal theta oscillations in early-stage Alzheimer disease. National Acad Sciences. Retrieved from http://www.pnas.org/content/early/2009/01/21/0811699106.abstract

37. Moran R, Zehetleitner M, Liesefeld HRR, Müller HJ, Usher M, Mueller HJ, Usher M (2016) Serial vs. parallel models of attention in visual search: accounting for benchmark RT-distributions. Psychon Bull Rev 23(5):1300–1315. https://doi.org/10.3758/s13423-015-0978-1

38. Parasuraman R, Greenwood PM, Alexander GE (2000) Alzheimer disease constricts the dynamic range of spatial attention in visual search. Neuropsychologia 38(8):1126–1135. https://doi.org/10.1016/S0028-3932(00)00024-5

39. Poil S-S, de Haan W, van der Flier WM, Mansvelder HD, Scheltens P, Linkenkaer-Hansen K (2013) Integrative EEG biomarkers predict progression to Alzheimer's disease at the MCI stage. Front Aging Neurosci 5:58. https://doi.org/10.3389/fnagi.2013.00058

40. Porter G, Leonards U, Wilcock G, Haworth J, Troscianko T, Tales A (2010a) New insights into feature and conjunction search: II. Evidence from Alzheimer's disease. Cortex 46(5):637–649. https://doi.org/10.1016/j.cortex.2009.04.014

41. Porter G, Tales A, Troscianko T, Wilcock G, Haworth J, Leonards U (2010b) New insights into feature and conjunction search: I. Evidence from pupil size, eye movements and ageing. Cortex. https://doi.org/10.1016/j.cortex.2009.04.013

42. Redel P, Bublak P, Sorg C, Kurz A, Förstl H, Müller HJ et al (2012) Deficits of spatial and task-related attentional selection in mild cognitive impairment and Alzheimer's disease. Neurobiol Aging 33(1). https://doi.org/10.1016/j.neurobiolaging.2010.05.014

43. Reiterer S, Pereda E, Bhattacharya J (2011) On a possible relationship between linguistic expertise and EEG gamma band phase synchrony. Front Psychol 2:334. https://doi.org/10.3389/fpsyg.2011.00334

44. Riddoch MJ, Chechlacz M, Mevorach C, Mavritsaki E, Allen H, Humphreys GW (2010) The neural mechanisms of visual selection: the view from neuropsychology. Ann N Y Acad Sci. https://doi.org/10.1111/j.1749-6632.2010.05448.x

45. Ryman DC, Acosta-Baena N, Aisen PS, Bird T, Danek A, Fox NC et al (2014) Symptom onset in autosomal dominant Alzheimer disease A systematic review and meta-analysis. Neurology 83(3):253–260. https://doi.org/10.1212/WNL.0000000000000596

46. Sparks DW, Proulx E, Lambe EK, Barrett K, Chattarji S (2018) Ready, set, go: the bridging of attention to action by acetylcholine in prefrontal cortex. J Physiol-Lond 596(9):1539–1540. https://doi.org/10.1113/JP275808

47. Stam C, Jones B, Nolte G, Breakspear M, Scheltens P (2006) Small-world networks and functional connectivity in Alzheimer's disease. Cereb Cortex 17(1):92–99. https://doi.org/10.1093/cercor/bhj127

48. Sultzer DL, Melrose RJ, Riskin-Jones H, Narvaez TA, Veliz J, Ando TK et al (2017) Cholinergic receptor binding in Alzheimer disease and healthy aging: assessment in vivo with positron emission tomography imaging. Am J Geriatr Psychiatr 25(4):342–353. https://doi.org/10.1016/j.jagp.2016.11.011

49. Tales A, Porter G (2008) Visual attention-related processing in Alzheimer's disease. Rev Clin Gerontol. https://doi.org/10.1017/S0959259809002792

50. Tales A, Butler SR, Fossey J, Gilchrist ID, Jones RW, Troscianko T (2002a) Visual search in Alzheimer's disease: a deficiency in processing conjunctions of features. Neuropsychologia 40(12):1849–1857. https://doi.org/10.1016/S0028-3932(02)00073-8

51. Tijms BM, Visser PJ (2018) Capturing the Alzheimer's disease pathological cascade. Lancet Neurol. Elsevier. https://doi.org/10.1016/S1474-4422(18)30043-7

52. Treisman AM, Gelade G (1980) Feature-integration theory of attention. Cogn Psychol 12(1):97–136. https://doi.org/10.1016/0010-0285(80)90005-5

53. Vallejo V, Cazzoli D, Rampa L, Zito GA, Feuerstein F, Gruber N et al (2016) Effects of Alzheimer's disease on visual target detection: a "peripheral bias". Front Aging Neurosci 8:1–8. https://doi.org/10.3389/fnagi.2016.00200

54. Watson DG, Humphreys GW (1997) Visual marking: prioritizing selection for new objects by top-down attentional inhibition of old objects. Psychol Rev 104(1):90–122. https://doi.org/10.1037/0033-295X.104.1.90

55. Whitehouse PJ, Martino AM, Antuono PG, Lowenstein PR, Coyle JT, Price DL, Kellar KJ (1986) Nicotinic acetylcholine binding sites in Alzheimer's disease. Brain Res 371(1):146–151. https://doi.org/10.1016/0006-8993(86)90819-X

56. Wolfe JM (2002) Visual search. Retrieved from http://search.bwh.harvard.edu/RECENT PROJECTS/visual_search_review/Review.html

57. Yener GG, Güntekin B, Öniz A, Başar E (2007) Increased frontal phase-locking of event-related theta oscillations in Alzheimer patients treated with cholinesterase inhibitors. Int J Psychophysiol 64(1):46–52. https://doi.org/10.1016/J.IJPSYCHO.2006.07.006

58. Yu AJ, Dayan P (2002) Acetylcholine in cortical inference. Neural Netw 15(4–6):719–730. https://doi.org/10.1016/S0893-6080(02)00058-8

59. Zamrini E, Maestu F, Pekkonen E, Funke M, Makela J, Riley M et al (2011) Magnetoencephalography as a putative biomarker for Alzheimer's disease. Int J Alzheimers Dis 2011:280289. https://doi.org/10.4061/2011/280289

Chapter 12
A Computational Hypothesis on How Serotonin Regulates Catecholamines in the Pathogenesis of Depressive Apathy

Massimo Silvetti, Gianluca Baldassarre, and Daniele Caligiore

Abstract Despite increasing literature supports a strong involvement of dopamine, noradrenaline and serotonin dysfunctions in the pathogenesis of most depressive disorders, the (causal) relationship between those monoamines impairments and the resulting disorder features is still not clear. We propose a hypothesis based on a computational model for which some depressive features may be produced by pathologically low levels of serotonin, which in turn causes a downregulation of catecholamine release. The simulations run with the model demonstrate that this process may be critical to the genesis of apathy, which is one of the most frequent and invalidating features of depressive disorders.

Keywords Apathy · Depression · Computational modeling · Dopamine · Noradrenaline · Serotonin

12.1 Introduction

Many of the symptoms seen in depression – such as anhedonia, amotivation, and apathy – have been consistently associated with dysfunctions in the catecholamines system ([12, 16, 27] for reviews). For example, experiments with rats exposed to chronic or unpredictable stressors have shown that there is an extended decrease in the activity of the dopamine (DA) neurons of the ventral tegmental area (VTA) [4, 33], resulting in reduced dopamine afflux to nucleus accumbens and lower inactivity ([19]; see Fiore et al. [10], for a computational model). Similarly, norepinephrine (Ne) activity in the locus coeruleus (LC) has been shown to be altered in patients with depression compared with controls, and this alteration influences the emer-

M. Silvetti
Department of Experimental Psychology, Ghent University, Ghent, Belgium
e-mail: massimo.silvetti@istc.cnr.it

G. Baldassarre · D. Caligiore (✉)
Institute of Cognitive Sciences and Technologies (ISTC-CNR), Rome, Italy
e-mail: gianluca.baldassarre@istc.cnr.it; daniele.caligiore@istc.cnr.it

© Springer Nature Switzerland AG 2019
V. Cutsuridis (ed.), *Multiscale Models of Brain Disorders*, Springer Series in
Cognitive and Neural Systems 13, https://doi.org/10.1007/978-3-030-18830-6_12

gence of symptoms like diminished ability to concentrate, depressed mood, and fatigue [3, 20, 25].

Aside from catecholamines, several studies investigating the physiopathology of depressive disorders have linked serotonin anomalies with the emergence of depressive symptoms [2, 6, 21, 37]. Although serotonin agonists represent a first-line treatment for depressive syndromes, the specific mechanisms through which they can affect depression symptoms are still elusive, probably due to the indirect and delayed mechanisms through which serotonin affects mood [14]. Important data supporting the involvement of serotonin in depression come also from studies of tryptophan depletion. In some works, a strong dietary manipulation is used to produce a transient lowering in brain serotonin activity through diminishing availability of its precursor amino acid, tryptophan [28]. It has also been shown that tryptophan depletion can imply depressive symptomatology in people who have experienced prior episodes of depression [22].

To date, it is not clear which is the relationship between the catecholamines dysregulation and the serotoninergic system in the pathogenesis of depressive symptoms [7]. In this chapter, we propose a hypothesis based on a computational model, for which low levels of serotonin disrupt optimal control of catecholamines release by altering the dialogue between the medial prefrontal cortex (MPFC) and the brainstem catecholamine nuclei. In particular, we hypothesize that pathologically low levels of serotonin translate into a higher internal cost of the release of catecholamines and hence to their downregulation. We propose that this process may contribute to the genesis of one of the most frequent and invalidating symptoms of the depressive syndrome: apathy.

12.2   The Reinforcement Meta-learner (RML) Model

To test our hypothesis, we used the RML model proposed in Silvetti et al. [31]. The RML architecture is based on the hypothesis that behavioral adaptation can be seen as a problem of decision-making [11, 23], whose objective is to maximize the long-term reward. In order to do so, animals need to learn both how to control optimally their behavior and how to control the internal variables that influence brain processes and behavior. For example, a foraging rat makes decisions about its movements in the environment. At the same time, it also needs to control higher level variables not directly linked to specific actions, like how much effort to invest (cost/benefit trade-off), whether to keep in consideration or not eventual changes in the surrounding environment (plasticity/stability trade-off), or even whether engaging or not in one activity. Learning to control optimally these internal variables is defined as "meta-learning" [26].

Many computational and theoretical studies indicate that the medial prefrontal cortex, and in particular the dorsal anterior cingulate cortex (dACC), is a multi-domain estimator of stimulus and action values that maximizes long-term reward (Reinforcement Learning (RL) view; [30]). RL computation by itself cannot fully account for dACC functioning as dACC is also linked to optimal control over internal variables that more broadly influence behavior (meta-learning; [1, 15, 29, 35]). The RML incorporates the hypothesis that it is critical to place dACC in a larger cortical-subcortical network to understand its elusive computational role and how the mammalian brain can perform optimal control over both behavior and the internal variables that regulate behavior (meta-learning perspective). At the neurophysiological level, the RML architecture is based on the demonstrated bidirectional anatomical connections between the dACC and the brainstem catecholamine nuclei [9, 17] – the VTA and the LC. As in earlier RL models, the RML computes the values of specific stimuli and actions to achieve adaptive behavior. However – and differently from earlier models – RML internal parameters are dynamically controlled by the interaction between the dACC and the brainstem catecholamine nuclei, implementing a *meta-learning* process.

The RML architecture is summarized in Fig. 12.1a. Here three loops of information flow are shown: an inner loop (between action selection and parameter control processes, black arrows); an external loop, between RML and environment (light gray arrows); and a third loop simulating RML control over other brain areas, via broadcasting catecholaminergic signal. The third loop shows how the dACC-brainstem system can work as a source of control signals to optimize the performance of other brain areas. An overview closer to neurophysiology (Fig.

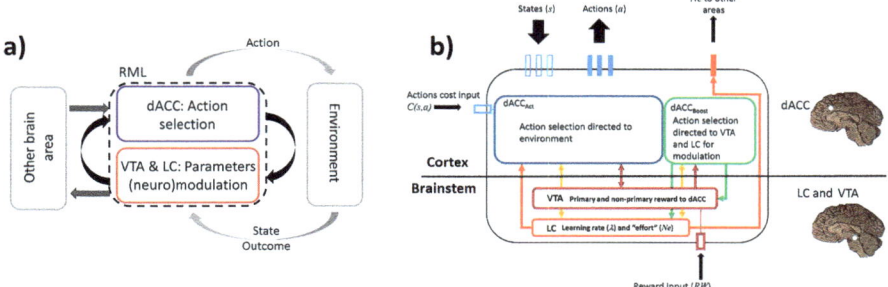

Fig. 12.1 (**a**) Model conceptual overview (see text). (**b**) Model overview with anatomical-functional analogy. The RML-environment interaction happens through nine channels of information exchange (black arrows) (input, empty bars; output, filled bars). The input channels consist of one channel encoding action costs (**c**), three encoding environmental states (s), and one encoding primary rewards (RW). The output consists of three channels each encoding one action (**a**), plus one channel conveying LC signals to other brain areas (Ne). The entire model is composed of four reciprocally connected modules (each with a different color). The upper modules (blue and green) simulate the dACC, while the lower modules (red and orange) simulate the brainstem catecholamine nuclei (VTA and LC). dACC$_{Act}$ selects actions directed toward the environment and learns through first and higher-order conditioning, whereas dACC$_{Boost}$ exploits similar computational principles to modulate catecholamine release. The VTA module provides DA training signals to both dACC modules, while the LC controls learning rate (yellow bidirectional arrow) in both dACC modules and effort exertion (promoting effortful actions) in the dACC$_{Act}$ module (orange arrow). Finally, the LC signal controlling effort in the dACC$_{Act}$ is directed also toward other brain areas for neuro-modulation. (See Silvetti et al. [31], for further details)

12.1b) shows that action selection and parameter control belong, respectively, to cortical and brainstem structures: the dACC (dACC$_{Act}$ and dACC$_{Boost}$ modules in Fig. 12.1b) performs action-outcome comparison and action selection while it is augmented by meta-learning (LC and VTA; orange/red modules in Fig. 12.1b). Both dACC$_{Act}$ and dACC$_{Boost}$ work in parallel and have the same objective of maximizing reward; nonetheless they differ not only for the target of their decisions (external environment for dACC$_{Act}$ and catecholamine nuclei for dACC$_{Boost}$) but also for the type of costs they have to keep in consideration for optimal decision-making. Indeed, while the dACC$_{Act}$ is influenced by action costs (e.g., energy expense to climb a ladder), the dACC$_{Boost}$ is influenced by the cost of enhancing catecholamine release. The latter is implemented as a linear function of the control signal afferent from the dACC$_{Boost}$ to the LC and VTA modules:

$$DA_B = r\,(R - \omega b) \tag{12.1}$$

where DA_B is the dopamine signal encoding reward for the dACC$_{Boost}$, r is a binary variable indicating whether the reward is present or not, R is the magnitude of the (eventual) reward, b is the control signal afferent from the dACC$_{Boost}$ itself, and ω is a parameter coding for the cost of boosting up catecholamine release. A complete detailed description of the RML model can be found in Silvetti et al. [31].

12.3 Modeling Serotonin Influence on Catecholamines Release

In the simulations described here, we hypothesize that serotonin plays a neuro-modulatory role on decision-making at a higher level with respect to catecholamines. We suggest that one of the serotonin functions consists in coding for the costs of enhancing (boosting) catecholamine release by the dACC [18]. In our model, this cost is represented by the parameter ω (Eq. 12.1). We hypothesize that low serotonin levels translate to higher ω values and hence in higher costs for catecholamine boosting. This means that, while catecholamines are responsible for optimal control of several decision-making variables (e.g., effort investment or plasticity), serotonin plays the higher-level role of regulating control over catecholamines by tuning the cost of their release.

12.4 Simulation Methods

We administered to the RML a decision-making task (here called "Effort Task") where effortful choices compete with low-effort choices (Fig. 12.2a; [24, 36]). This task consists in a sequence of trials where binary choices must be made. One of the choices requires a high effort to be executed and leads to high reward, while the

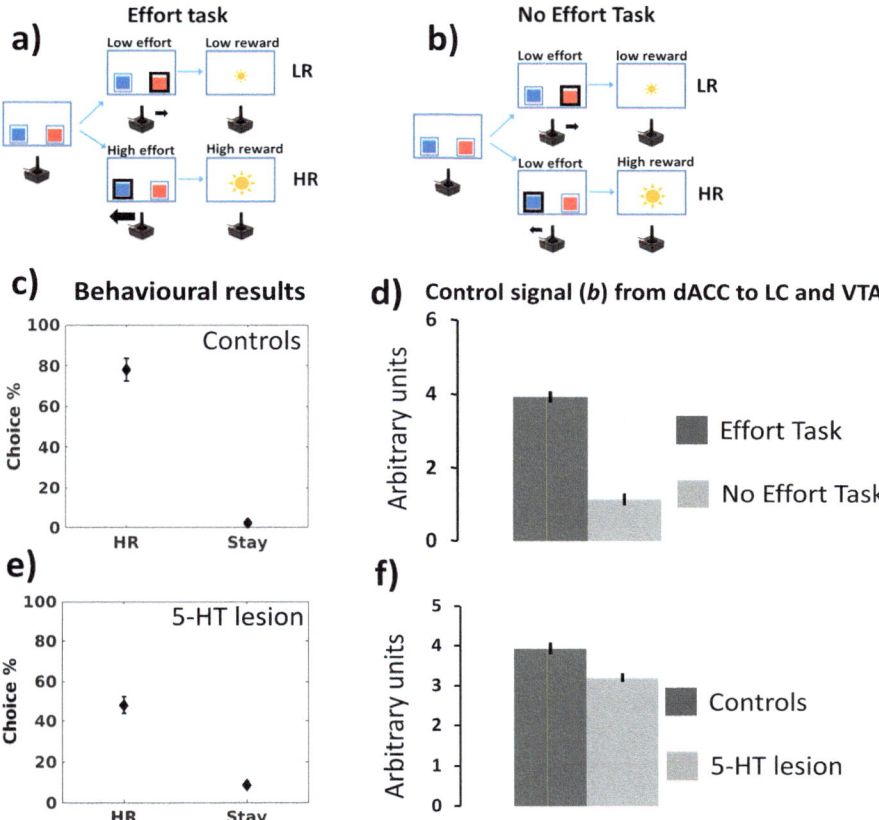

Fig. 12.2 (**a, b**) Effort and No Effort tasks executed by the RML. Here action selection is represented by a black arrow that selects either a blue or a red box. The effort needed to execute the action is proportional to the arrow thickness. (**c, e**) Behavioral results, respectively, in controls and 5-HT lesioned (simulated) subjects. (**d, f**) Control signal selected by the dACC$_{Boost}$ to modulate LC and VTA activity

other one implies a low effort leading to small reward. We simulated two groups, one control group (where all the RML parameters are the same as in Silvetti et al. [31]) and one group simulating serotonin depletion. The latter was implemented through a 100% increase of the ω parameter in Eq. 12.1. Moreover, we administered to the control group also a "No Effort Task" where both the choices required a low effort, but one kept on leading to a higher reward (Fig. 12.2b). To mimic standard experimental paradigms, we repeated each simulation 12 times (i.e., 12 simulated subjects for each group). This verified that the model could generate a large effect size of the results (p-values can always be improved by running more simulated subjects).

12.5   Results and Discussion

The RML simulating the control group prefers to select the high-effort-high-reward option (Fig. 12.2c), in agreement with experimental data from Walton et al. [36]. At the neurophysiological level (Fig. 12.2d), the $dACC_{Boost}$ increased the boosting signal (b in Eq. 12.1) toward the catecholamines nuclei (VTA and LC) in the Effort task compared to the No Effort Task ($t(11) = 18.78$, $p < 0.0001$), simulating the experimental results with nonhuman primates [34]. Increased catecholamines influence decision-making in the $dACC_{Act}$ by energizing behavior and facilitating effortful actions (Fig. 12.2c). At the same time, boosting catecholamines has a cost (Eq. 12.1), so that the higher the variable b, the higher the reward discount for the $dACC_{Boost}$ module. The result of these two opposite forces (energizing behavior by catecholamines boosting and minimizing the cost of boosting itself) converges to the optimal value of b and therefore of catecholamine release by VTA and LC. In case of low serotonin levels (Fig. 12.2e), the RML exhibits a behavior with higher selection percentage of low effort-low reward choice. Moreover, serotonin depleted RML shows also a higher percentage of refusals to engage in the task (Stay option). This behavioral pattern indicates a lowered capability to exert effort for achieving a reward, with both the tendency to minimize effort rather than to maximize reward and a higher probability of immobility (refusal to engage in the task). This behavioral impairment closely simulates behavioral apathy in both human patients and animal models. At the neurophysiological level, the RML predicts a downregulation of catecholamine release (Fig. 12.2f; $t(11) = 2.98$, $p = 0.013$) in agreement with experimental data from depressed patients [12, 16]. In summary, the final consequence of low serotonin levels is a downregulation of catecholamine release because the dACC reward-based decision-making processing finds less valuable to boost up catecholamines.

12.6   Conclusions

The role of serotonin in depression is not clear yet. It has been shown that impairing serotonin function can cause clinical depression only in some circumstances, for example, in patients with high-risk factors for depression [22, 32]. Based on these data, it has been suggested that low serotonin function may compromise mechanisms involved in maintaining recovery from depression rather than having a primary effect in symptoms generation [7]. In this chapter, we propose a computational model that starts to address this issue by making explicit a possible neural mechanism underlying the involvement of serotonin in the emergence of depressive apathy. In particular, simulation results suggest that low levels of serotonin may cause an increased evaluation of costs about catecholamine release control, with consequent catecholamine downregulation, leading to apathy. This hypothesis might be also linked to the strong interaction of serotonin with amygdala, important for the appraisal of negative experiences [13, 14]. Moreover, it is also in line with influent theoretical/computational modelling proposals supporting the role of serotonin in cost evaluation [5, 8]. These data represent the first step of a research agenda aiming at understanding the link between neuromodulators and computational aspects of decision-making in the pathogenesis of depressive symptoms.

References

1. Behrens TE, Woolrich MW, Walton ME, Rushworth MF (2007) Learning the value of information in an uncertain world. Nat Neurosci 10(9):1214–1221
2. Carlssen A (1976) The contribution of drug research to investigating the nature of endogenous depression. Pharmacopsychiatry 9(01):2–10
3. Chamberlain SR, Robbins TW (2013) Noradrenergic modulation of cognition: therapeutic implications. J Psychopharmacol 27(8):694–718
4. Chang CH, Grace AA (2014) Amygdala-ventral pallidum pathway decreases dopamine activity after chronic mild stress in rats. Biol Psychiatry 76(3):223–230
5. Cools R, Nakamura K, Daw ND (2011) Serotonin and dopamine: unifying affective, activational, and decision functions. Neuropsychopharmacology 36(1):98–113
6. Coppen A (1967) The biochemistry of affective disorders. Br J Psychiatry 113(504):1237–1264
7. Cowen PJ, Browning M (2015) What has serotonin to do with depression? World Psychiatry 14(2):158–160
8. Daw ND, Kakade S, Dayan P (2002) Opponent interactions between serotonin and dopamine. Neural Netw 15(4):603–616
9. Devinsky O, Morrell MJ, Vogt BA (1995) Contributions of anterior cingulate cortex to behaviour. Brain 118(Pt 1):279–306
10. Fiore VG, Mannella F, Mirolli M, Latagliata EC, Valzania A, Cabib S, Dolan RJ, Puglisi-Allegra S, Baldassarre G (2015) Corticolimbic catecholamines in stress: a computational model of the appraisal of controllability. Brain Struct Funct 220:1339–1353
11. Frank MJ, Seeberger LC, O'Reilly RC (2004) By carrot or by stick: cognitive reinforcement learning in parkinsonism. Science 306(5703):1940–1943
12. Grace AA (2016) Dysregulation of the dopamine system in the pathophysiology of schizophrenia and depression. Nat Rev Neurosci 17(8):524
13. Haas BW, Omura K, Constable RT, Canli T (2007) Emotional conflict and neuroticism: personality-dependent activation in the amygdala and subgenual anterior cingulate. Behav Neurosci 121(2):249
14. Harmer CJ, Goodwin GM, Cowen PJ (2009) Why do antidepressants take so long to work? A cognitive neuropsychological model of antidepressant drug action. Br J Psychiatry 195(2):102–108
15. Kolling N, Wittmann MK, Behrens TEJ, Boorman ED, Mars RB, Rushworth MFS (2016) Value, search, persistence and model updating in anterior cingulate cortex. Nat Neurosci 19(10):1280–1285
16. Maletic V, Eramo A, Gwin K, Offord SJ, Duffy RA (2017) The role of norepinephrine and its α-adrenergic receptors in the pathophysiology and treatment of major depressive disorder and schizophrenia: a systematic review. Front Psych 8:42
17. Margulies DS, Kelly AMC, Uddin LQ, Biswal BB, Castellanos FX, Milham MP (2007) Mapping the functional connectivity of anterior cingulate cortex. NeuroImage 37(2):579–588
18. Meyniel F, Goodwin GM, Deakin JW, Klinge C, MacFadyen C, Milligan H et al (2016) A specific role for serotonin in overcoming effort cost. elife 5:e17282
19. Pascucci T, Ventura R, Latagliata EC, Cabib S, Puglisi-Allegra S (2007) The medial prefrontal cortex determines the accumbens dopamine response to stress through the opposing influences of norepinephrine and dopamine. Cereb Cortex 17:2796–2804

20. Rajkowska G (2000) Histopathology of the prefrontal cortex in major depression: what does it tell us about dysfunctional monoaminergic circuits? Prog Brain Res 126:397–412
21. Ressler KJ, Nemeroff CB (2000) Role of serotonergic and noradrenergic systems in the pathophysiology of depression and anxiety disorders. Depress Anxiety 12(S1):2–19
22. Ruhé HG, Mason NS, Schene AH (2007) Mood is indirectly related to serotonin, norepinephrine and dopamine levels in humans: a meta-analysis of monoamine depletion studies. Mol Psychiatry 12:331–359
23. Rushworth MF, Behrens TE (2008) Choice, uncertainty and value in prefrontal and cingulate cortex. Nat Neurosci 11(4):389–397
24. Salamone JD, Cousins MS, Bucher S (1994) Anhedonia or anergia? Effects of haloperidol and nucleus accumbens dopamine depletion on instrumental response selection in a T-maze cost/benefit procedure. Behav Brain Res 65(2):221–229
25. Saltiel PF, Silvershein DI (2015) Major depressive disorder: mechanism-based prescribing for personalized medicine. Neuropsychiatr Dis Treat 11:875
26. Schweighofer N, Doya K (2003) Meta-learning in reinforcement learning. Neural Netw 16(1):5–9
27. Schildkraut JJ (1965) The catecholamine hypothesis of affective disorders: a review of supporting evidence. Am J Psychiatr 122(5):509–522
28. Sharp T, Cowen PJ (2011) 5-HT and depression: is the glass half-full? Curr Opin Pharmacol 11(1):45–51
29. Shenhav A, Botvinick MM, Cohen JD (2013) The expected value of control: an integrative theory of anterior cingulate cortex function. Neuron 79(2):217–240
30. Silvetti M, Alexander W, Verguts T, Brown JW (2014) From conflict management to reward-based decision making: actors and critics in primate medial frontal cortex. Neurosci Biobehav Rev 46:44–57
31. Silvetti M, Vassena E, Abrahamse E, Verguts T (2017) Dorsal anterior cingulate-midbrain ensemble as a reinforcement meta-learner. *bioRxiv*:130195
32. Smith KA, Fairburn CG, Cowen PJ (1997) Relapse of depression after rapid depletion of tryptophan. Lancet 349(9056):915–919
33. Valenti O, Gill KM, Grace AA (2012) Different stressors produce excitation or inhibition of mesolimbic dopamine neuron activity: response alteration by stress pre-exposure. Eur J Neurosci 35(8):1312–1321
34. Varazzani C, San-Galli A, Gilardeau S, Bouret S (2015) Noradrenaline and dopamine neurons in the reward/effort trade-off: a direct electrophysiological comparison in behaving monkeys. J Neurosci 35(20):7866–7877
35. Verguts T, Vassena E, Silvetti M (2015) Adaptive effort investment in cognitive and physical tasks: a neurocomputational model. Front Behav Neurosci 9:57
36. Walton ME, Groves J, Jennings KA, Croxson PL, Sharp T, Rushworth MF, Bannerman DM (2009) Comparing the role of the anterior cingulate cortex and 6-hydroxydopamine nucleus accumbens lesions on operant effort-based decision making. Eur J Neurosci 29(8):1678–1691
37. Yohn CN, Gergues MM, Samuels BA (2017) The role of 5-HT receptors in depression. Mol Brain 10(1):28

Chapter 13
Autism Spectrum Disorder and Deep Attractors in Neurodynamics

Włodzisław Duch

Abstract Behavior may be analyzed at many levels, from genes to psychological constructs characterizing mental events. Neurodynamics is at the middle level. It can be related to biophysical properties of neurons that depend on lower-level molecular properties and genetics and used to characterize high-level processes correlated with behavior and mental states. A good strategy that should help to find causal relations between different levels of analysis, showing how psychological constructs used in neuropsychiatry emerge from biology, is to identify biophysical parameters of neurons required for normal neural network activity and explore all changes that may lead to abnormal functions, behavioral symptoms, cognitive phenotypes, and psychiatric syndromes. Neural network computational simulations, as well as analysis of real brain signals, show importance of attractor states, providing language that can be used to explain many features of mental disorders. Computational simulations of neurodynamics may generate hypothesis for experimental verification and help to create mechanistic explanation of observed behavior. Autism spectrum disorder is used as an example of the usefulness of such approach, showing how deep attractors resulting from ion channel dysfunctions slow down attention shifts, influence connectivity, and lead to diverse developmental problems.

Keywords Neurodynamics · RDoC · Mental disorders · Autism spectrum disorder (ASD) · Brain fingerprints · Computational modeling

13.1   Neurodynamics and Many Levels of Neuropsychiatry

Diagnostic criteria at the foundation of psychiatry and clinical psychology contained in the *Diagnostic and Statistical Manual of Mental Disorders* [6] are based on

W. Duch (✉)
Department of Informatics, Faculty of Physics, Astronomy and Informatics, and Neurocognitive Laboratory, Center for Modern Interdisciplinary Technologies, Nicolaus Copernicus University, Toruń, Poland
e-mail: wduch@is.umk.pl

evaluation of behavioral symptoms. Research Domain Criteria (RDoC) is an attempt by NIMH to integrate many levels of information needed to understand human behavior [2, 3]. Psychological constructs are used to characterize five general domains: arousal and regulatory systems, negative and positive valence systems, cognitive systems, and social processes. Many psychological constructs and more detailed subconstructs are used for each domain, and each construct is described by "units of analysis" that include specific genes, molecules, cells, circuits, physiology, behavior, self-reports representing psychological components, and paradigms defining experimental procedures. The RDoC matrix based on constructs vs. unit analysis is far from being complete and is not yet useful to build models of functions based on activity of brain subnetworks. In particular it does not characterize different types of neurons in terms of their structure, synapses, receptors, ion channels, connectivity, and other "units of analysis" that influence network functions.

Although all RDoC units of analysis are important, understanding the mechanics of mental functions should be done at the circuit level. Functions of neural networks depend on the cellular, molecular, and genetic levels. Complex functions responsible for behavior result from neurodynamics. Therefore a good strategy that should help to find causal relations between different levels of analysis, showing how RDoC psychological constructs emerge from biology, is to identify biophysical parameters of neurons required for normal neural network activity and explore all changes that may lead to abnormal functions, behavioral symptoms, cognitive phenotypes, and syndromes. Computational simulations of neurodynamics generate hypothesis for experimental verification and help to interpret neuroimaging data. Neurodynamics provides language that relates measureable brain processes to RDoC psychological constructs. As an example of such an approach, I shall focus here on the autism spectrum disorders (ASD). Many confusing observations may find an explanation at this level and lead to hypothesis that may be experimentally verified.

13.2 Attempts to Understand Autism Spectrum Disorders

There is a growing consensus that autism is not a single disease but belongs to a spectrum of various disorders of general temporospatial neural processing [14]. Many specific mechanisms and multiple etiologies causing ASD may exist, including metabolic and immune system deregulation, exposure to various chemicals, and other environmental factors [29]. Based on the DSM criteria, core behavioral symptoms may be sufficient for the diagnosis of autism, but RDoC characterization will reveal phenotypic diversity, with each subgroup requiring different approach to therapy. Many brain diseases (ASD, spectrum of psychotic disorders, epilepsy) should be placed in a continuum phenomics space, forming a spectrum of diseases that may have similar core symptoms, but great variability of all RDoC units of analysis. In the case of ASD, even the main symptoms are highly variable. There are many theories of autism that focus on selected aspects of behavior or clinical observations, mistaking symptoms for deeper causes [30].

So far big projects related to autism have been focused mostly on a single level. A lot of efforts have been devoted to the genetics of autism. Large number of genes involved in deregulation of neural systems has clearly shown that major brain diseases may have very different etiologies. Research on the genetics of autism has identified over 880 genes (about 4.5% of all human genes) that are correlated with some form of autism (SFARI Gene database, Q1 2017 release, https://gene.sfari.org/). They are involved in cell signaling, structure and transport, metabolic, immune and neural processes, and frequently implicated in other disease such as cancer, cardiac or neurodegenerative disease [29]. Genetic variation and environmental conditions lead to the diversity of proteins, signaling pathways, ion channels, synapses, and structures of neurons and their connections. Unfortunately there are no good methods to analyze in vivo molecular structure of biological neurons.

The motto of molecular biology "structure is function" is also true at the systems level. Therefore the best strategy is to analyze neural properties in relation to molecular and genetic levels and investigate how that will influence neurodynamics, spatiotemporal patterns of neuronal electrical activity. Brain functions and observable behavioral symptoms may then be understood in terms of specific dysfunctions of neurons. To achieve this goal, the whole causal chain (Fig. 13.1) sketched below is needed.

1. Genes are expressed in different parts of the brain, creating proteins that form neural receptors, ion channels, synapses, and cell membranes. Mutations, copy number variation, and other genetic processes create specific dysfunctions of proteins building ion channels, influencing generation of action potentials [22].
2. Complete ion channelome is needed and should be related to different types of neurons, their dendrites, axons and membranes, and density and distributions of ion channels, influencing integration of synaptic inputs [12].
3. Specific character of individual neurons depends on all biophysical properties, but the distribution and temporal activation of voltage-gated ion channels are of particular importance. The fast temporal dynamics of activity-driven ion channel changes should be taken into account [18].
4. Neural simulators aimed at detailed modeling of single neurons at subcellular component level, including biochemical reactions, are needed to investigate how changes at molecular level determine properties of single neurons and how these properties influence, in stimulus-driven situations, development of neural networks and whole connectomes. Such neural simulators are in the early stage of development. NEURON, GENESIS 3 and the hope is that Brain Simulation

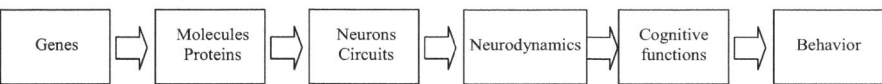

Fig. 13.1 Causal chain for understanding of ASD mechanics

Platform of the Human Brain Project will provide even more detailed simulators that should integrate all experimental information.

5. Simulations of brain functions related to the five RDoC domains should reveal the range of biophysical neural parameters that may be responsible for normal functions and disruption of these functions. Whole-brain simulations should also show how connectomes develop as a function of sensory stimulation and internal dynamics.

A lot of data is missing at each of these stages. Understanding this causal chain is a real challenge in ASD research, and it should be clearly stated as a vision based on RDoC approach and one of the goals of the HBP.

Although the complexity of the problem is overwhelming as a kind of a "proof of principle," I shall show below how multilevel approach may be applied to ASD, generating hypothesis that may be experimentally verified.

13.3   ASD and Neurodynamics

As the first approximation, minimal models that capture some properties of biological networks and allow for simulation of experimental observations are needed. Starting from simple models of neurons and networks, we have tried to create models of normal cognitive and motor functions and determine ranges of model parameters that preserve these functions. Synchronization of neurons in local microcircuits and between distal brain areas is necessary for binding neuronal activations that permit perception, action, and other cognitive activity. Abnormal temporospatial neural processing [14] is at the root of pervasive developmental disorders but also attention deficit disorder, both in the inattentive and hyperactive (ADHD) form, concentration deficit disorder, bradyphrenia (slowness of thought), and other disorders related to attention. Such effects may be investigated using attractor neural networks [1], where the activity of groups of neurons settles in a quasi-stable spatiotemporal pattern called "attractor." These patterns encode long-term memory, concepts, and object recognition. The subspace of initial activations that will end in the attractor as a result of neural dynamics is called "the basin of attractor." Transitions between attractors are possible due to the noise in the system, effects of neural fatigue, or signals coming from other groups of neurons due to the external or internal stimulation.

Neurodynamics takes place at many spatial and time scales, from the nanoscale to slow developmental and learning (neuroplasticity) processes. Relevance of these processes depends on the questions that are asked. In analogy to adiabatic approximation in quantum systems, one can consider transitions in neurodynamics as relatively independent of slower processes responsible for neuroplasticity. In this approximation neurodynamics may be investigated on a train network that has already fixed synaptic connections and may assume many distinct attractor states. However, one should remember that the development of connectomes due to the Hebbian associative mechanisms depends on the stimulations that create attractor states in networks. Attractors in the sensory cortices develop quite early in infancy; perception-action cycle attractors develop later, coupling local attractor states, synchronizing the activity in sensory and motor areas through distal connections [27]. Formation of such attractors depends on frequency of stimulations and the time the system stays in a given state and induces neuroplastic changes. The time in which neurodynamics dwells in a given attractor basin should be within certain range to ensure normal development. If local attractors are too strong, capturing neurodynamics for a long time, the effective number of internal changes of activation patterns may be low. This will prevent formation of stronger and more complex attractors connecting wider brain areas and thus lead to the underconnectivity between distant brain areas.

The underconnectivity theory of autism has achieved considerable success [19], but reasons for local overconnectivity and underconnectivity (or lower bandwidth of information transmission) between frontal and posterior brain areas need deeper explanation that may be provided by the deep attractor theory. Zimmerman book [30] describes 20 different approaches to ASD, divided into 6 types: molecular and clinical genetics; neurotransmitters and cell signaling; endocrinology, growth, and metabolism; immunology, maternal-fetal interaction, and neuroinflammation; environmental mechanisms and models; and neuroanatomy and neural networks. Most approaches focus on phenomenological observations. Minicolumnopathy, mirror neuron system (MNS), theory of mind, underconnectivity, empathizing-systemizing, and executive dysfunction theory all focus on symptoms trying to link them to behavior. Such approaches do not provide an explanation why such symptoms arise and why observed abnormalities create specific behavioral problems. Imbalanced spectrally timed adaptive resonance theory [16], or iSTART, is based on artificial neural network that does not include measurable parameters. This model simply assumes breakdown of some brain functions – underaroused emotional depression, hyperspecific learning, and attentional and motor circuits – but has no relations to the biophysical reality at molecular or neuroimaging levels.

Neurodynamics depends on many parameters that characterize neurons and their networks: general network connectivity; types of neurons; density and strength of synaptic coupling; the balance between excitatory, inhibitory, and leak currents; types of ion channels (ligand or voltage-gated, inward-rectifier); availability of neurotransmitters; and many others. Construction of computational models incorporating all details is not yet feasible, but even greatly simplified models may help to generate useful insight into some brain functions.

13.4 Computational Simulations

Minimal model of neurons that can be linked to biophysical reality should include excitatory and inhibitory ion channels and leak channels that control spontaneous depolarization. Emergent neural simulation software based on Leabra cognitive

Fig. 13.2 Point neuron, three
types of ion channels

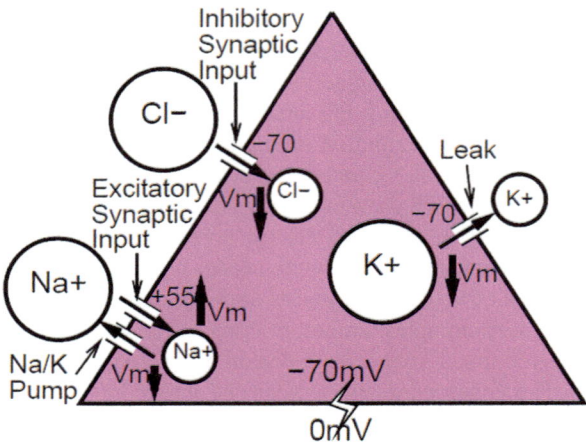

architecture is relatively simple and captures most important biological principles
[24, 25]. Point neurons and rate coding of neural activity is used to replace
population of spiking neurons by single units, three types of ion channels (Fig.
13.2), k-winners-takes-all (kWTA) mechanism to account for inhibition and sparse
coding, several types of noise, and Hebbian and error-driven learning mechanisms.
This architecture has been developed over several decades and is implemented in
the Emergent neural simulation software, providing a great tool for a whole family
of simple attractor network models of various brain functions that may be used to
illustrate under which conditions normal functions are disrupted.

I will summarize here three types of models relevant to autism that we have
investigated in the past: attention shifts [9, 11, 15], spontaneous thought dynamics
[8, 10] based on the model of reading, and simple cyclic movements [7].

The attention shift model has been based on classical Posner spatial cueing task.
Model implemented in Emergent (Fig. 13.3) is composed of input, V1, and two
spatial and two object recognition layers with additional output layer. This model
is essentially the same as described in O'Reilly and Munakata [25], simulating
the speed of reaction times when helpful (valid) or confusing (invalid) cues are
presented. Speedup and slowdown for valid/invalid trials compared to a neutral trial
with no cueing can be calculated. Effects of lesions in case of hemispatial neglect
and Balint's syndrome have also been shown in this model, showing significantly
slower reaction times in case of invalid cues.

Problems with the speed of attention shifts may arise not only due to the lesions
but also changes in relative strength of excitatory/inhibitory and leak ion channel
conductances. In many investigations individuals with ASD have shown atypical
attention patterns. For example, Landry and Bryson [23] found that "Children
with autism had marked difficulty in disengaging attention. Indeed, on 20% of
trials they remained fixated on the first of two competing stimuli for the entire 8-
second trial duration." Kawakubo et al. [20] conclude: "We suggest that adults with
autism have deficits in attentional disengagement and the physiological substrates

Fig. 13.3 Posner spatial
cueing task model
implemented in Emergent
simulator

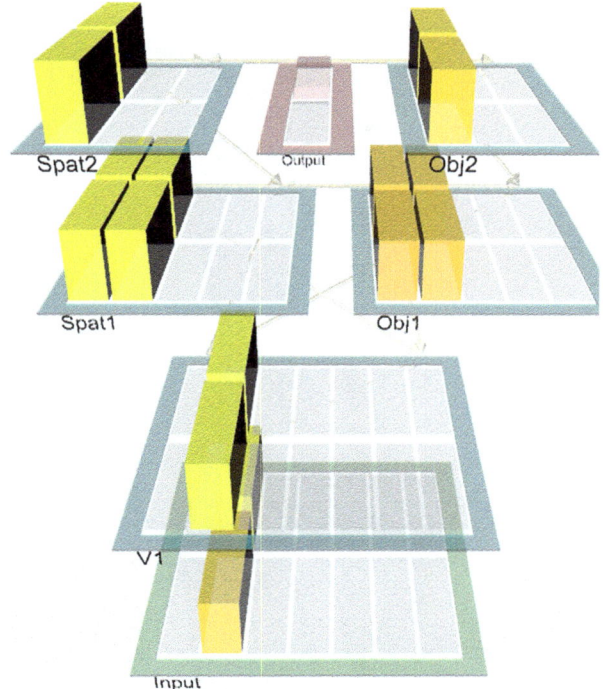

underlying deficits in autism and mental retardation are different. [...] These
results demonstrate electrophysiological abnormalities of disengagement during
visuospatial attention in adults with autism which cannot be attributed to their IQs."
Development of such problems is gradual – between 7- and 14-month infants who
were later diagnosed with autism stopped improving speed and flexibility of their
visual orientation [13].

Our simulations of attention shift effects point to the mechanism that is also
seen in spontaneous transition between thoughts and cyclic movements in case of
motor system activations. The model of thought wandering is based on a modified
model of normal reading and dyslexia, implemented in the Emergent simulator
[25]. The model has six layers, representing information about orthography (6 × 8
units), phonology (14 × 14 units), and semantics (10 × 14 units), connected to
each other via intermediate (hidden) layers of neurons (Fig. 13.4). Full connectivity
between each adjacent layer is assumed, with recurrent self-connections within each
of these layers. The original model has been used primarily to study various forms of
dyslexia due to the lesions of one of the intermediate layers between the two inputs
and the semantic layer. The network has been trained on 40 words, half of them con-
crete and half of them abstract. Semantics has been captured by using micro-features
describing words. Accommodation mechanism has been added, based on the concentration of intracellular calcium that builds up slowly as a function of activation

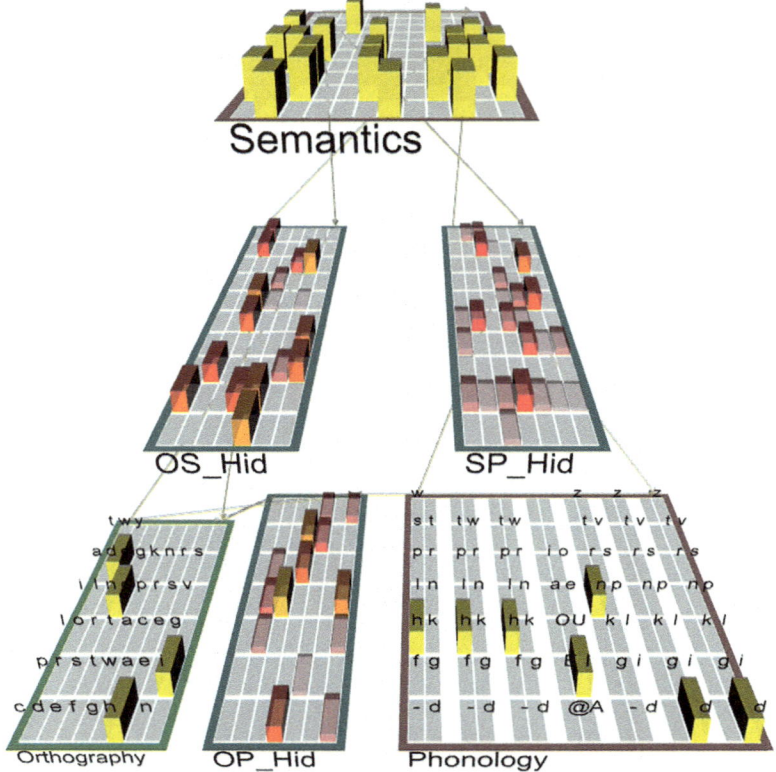

Fig. 13.4 Model of reading, with orthographic, phonological and semantic layers (140 units representing microfeatures) and 3 intermedite layers, showing activity of neurons during mind-wondering

and opens leak channels releasing potassium ions, regulating subsequent inhibition of a neuron. Synaptic Gaussian noise with zero mean and 0.02 variance has been used to facilitate free transitions between attractors representing words or thoughts. The network is prompted by showing it a word in the orthographic or phonological layer and observing transitions of activity in the semantic layer neurons.

To see trajectories of neurodynamics in 140-dimensional space, recurrence plots (RPs) and fuzzy symbolic dynamics (FSD) visualization have been used [8, 10]. In Fig. 13.5 examples of such trajectories are shown for three values of parameter controlling the calcium buildup: $b = 0.005$ leads to deep attractor basins and reduced number of states in neurodynamics, $b = 0.01$ leads to normal transitions, and $b = 0.02$ leads to fast depolarization of neurons, shallow attractor basins, and the inability to dwell in a single state. In the first case, neurons remain synchronized in one persistent pattern; trajectories of neurodynamics are trapped in attractor basins

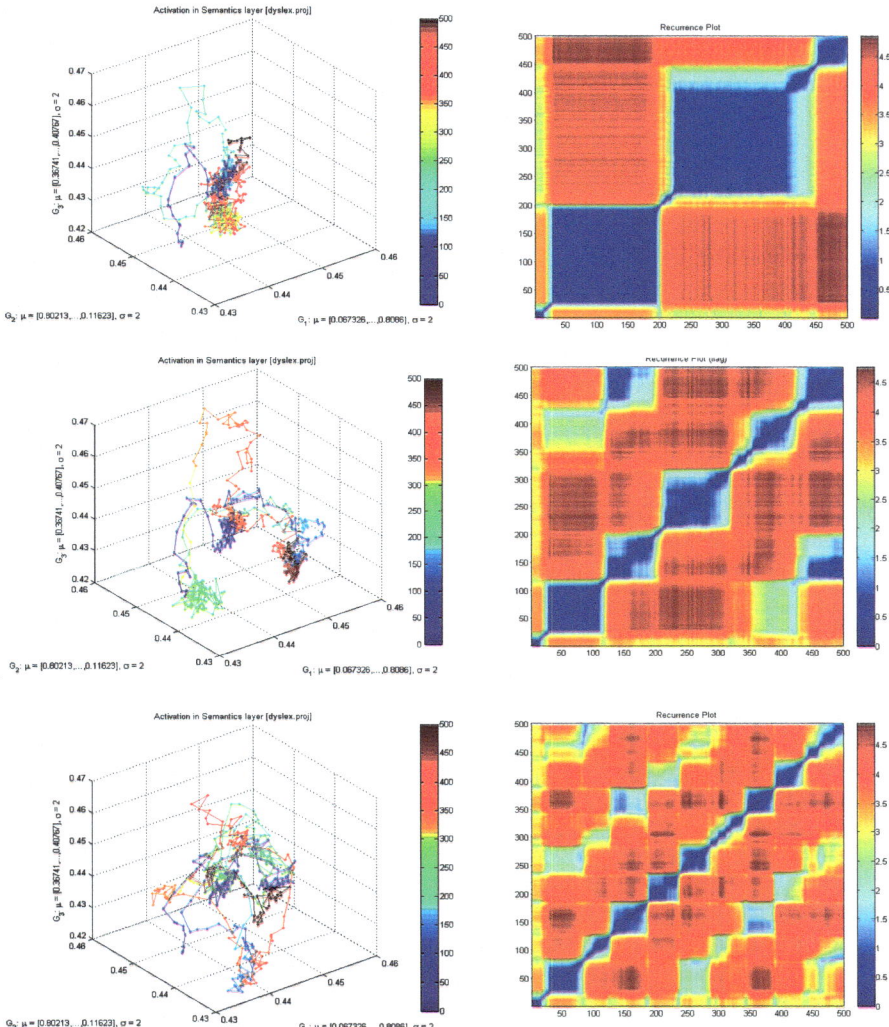

Fig. 13.5 FSD (left) and RP (right) visualization of attractors in semantic layer with 140 units in weak (ASD), normal and strong (ADHD) accommodation case

for relatively long time. This seems to explain why disengagement of attention in ASD is slow.

On the other hand, too short synchronization times, or shallow basins of attractors, lead to rapid jumps from one basin of attraction to another, with short dwell times. Attention is not focused long enough, as is typical in case of attention deficit hyperactivity disorder (ADHD). Thus a single parameter that controls neural accommodation mechanism may lead to very different behaviors. Some ASD and

ADHD cases may be at the opposite ends of the same spectrum. Other mechanisms (inhibition, recurrence) may be lead to different subtypes of these diseases.

Links between calcium and potassium channelopathies and ASD have been recently noticed [17]. Several genes (CACNA1C, CACN1G, and CACNA1I) that control construction of calcium ion channels have been associated with ASD.

13.5 Interpretation

Strong synchronization of neurons in local sensory cortices creates deep attractors trapping neurodynamics for a long time. Network activity patterns do not change with normal frequency, and therefore perception-action networks requiring synchronization of distant cortical areas are developing slowly. This is consistent with many observations related to the development of frontoparietal connectivity [19]. Courchesne and Pierce [5] have also postulated that ASD is characterized by early local hyperconnectivity and a long-distance hypoconnectivity of the prefrontal cortex. Trying to understand conflicting neuroimaging findings of hypo- and hyperconnectivity in children and adults, Uddin et al. [28] suggested that the increase in functional connectivity over the age span may be slower in ASD group. Deep attractor hypothesis supports these views and links them to properties of neurons at molecular and genetic levels. Analysis of neurodynamics has great diagnostic potential and may be used after EEG source power reconstruction or after wavelet multiscale decomposition of signal. Recurrence quantitative analysis estimating trapping time, recurrence rate, entropy, laminarity, and other nonlinear features of EEG signal allowed for discrimination between autistic and typically developing children starting at 3 months of age [4].

In our computational models, strong attractors may arise due to several reasons: unusually strong inhibition, strong recurrence, or damage of leak (K+) ion channels that has genetic basis. Shift of attention due to the bottom-up processes in Posner experiment requires desynchronization of current activation patterns and resynchronization of the new one. Spontaneous depolarization of neurons through the leak ion channels plays an important role in this process. Sizes of basins of attractors may considerably differ depending on encoding of stimulus and how initial connectivity was structured. Hyperconnectivity may lead to relatively small but very strong basins. One way to estimate it in case of attractor network is to plot variance of the fluctuations $\sigma(P(\varepsilon))$ around the mean attractor pattern P as a function of the synaptic noise ε. If the variance is initially low for growing noise variance, but at some point there is a sharp increase, the attractor basin is deep (synchronization is strong) and narrow (fluctuations are small). Behavioral interpretation of such situation is that even strong stimuli will be ignored, resulting in underreaction. Deep attractors may be activated in the cortex even when sensory stimulation is rather weak, and this may be true for all sensory modalities, sight, hearing, touch, smell, movement, and taste but also purely internal activation. From behavioral perspective deep attractors in perception-action cycle will lead to insistence on sameness. Development of strong attractors coupling sensory cortices and subcortical areas controlling emotions may result in overreaction and tantrums [26].

On the other hand, if the variance $\sigma(P(\varepsilon))$ of the network activation patterns will grow with increasing noise, the attractor basin may be broad, shifts of attentions may be easier, and development of long-distance connections should be faster. To achieve such desirable outcome, children should be stimulated in an intensive way. Applied behavioral analysis is using such intensive stimulation and is the best-established form of therapy for children with autism. Detailed simulations of trajectories entering basins of attractors show that steps of the trajectories (total change of patterns in short time step) decrease near the center, making it hard in case of ASD to get out of the basin of attractor. A flow of activity may prevent the tendency to dwell for longer time in one state (perception, thought, action). For example, using the rapid sequential visual presentation technique, one can adjust speed that allows for comprehension but does not allow to maintain the same brain state for long.

Finding fingerprints of persistent EEG activity in brains of autistic children should give support to ideas presented here. Many other ideas may be derived from computational simulations and the deep attractor hypothesis. More detailed computational simulations should help to understand casual chain linking genetics, neural structures, development of connectomes, and behavior in a meaningful way. The language of dynamical systems may help to bridge the gap between physical and mental processes.

Acknowledgments This research was supported by the National Science Center, Poland, UMO-2016/20/W/NZ4/00354. Visualizations of trajectories have been made using VISER Toolbox developed by Krzysztof Dobosz in our laboratory.

References

1. Amit DJ (1992) Modeling brain function: the world of attractor neural networks. Cambridge University Press, Cambridge
2. Bilder RM, Sabb FW, Cannon TD, London ED, Jentsch JD, Parker DS et al (2009a) Phenomics: the systematic study of phenotypes on a genome-wide scale. Neuroscience 164(1):30–42
3. Bilder RM, Sabb FW, Parker DS, Kalar D, Chu WW, Fox J et al (2009b) Cognitive ontologies for neuropsychiatric phenomics research. Cogn Neuropsychiatry 14(4–5):419–450
4. Bosl WJ, Tager-Flusberg H, Nelson CA (2018) EEG analytics for early detection of autism Spectrum disorder: a data-driven approach. Sci Rep 8(1):6828
5. Courchesne E, Pierce K (2005) Why the frontal cortex in autism might be talking only to itself: local over-connectivity but long-distance disconnection. Curr Opin Neurobiol 15:225–230
6. Diagnostic & Statistical Manual of Mental Disorders (2013, 5th ed) American Psychiatric Association, Washington, DC
7. Dobosz K, Mikołajewski D, Wójcik GM, Duch W (2013) Simple cyclic movements as a distinct autism feature-computational approach. Comput Sci 14(3):475–489
8. Dobosz K, Duch W (2010) Understanding neurodynamical systems via fuzzy symbolic dynamics. Neural Netw 23:487–496

9. Duch W, Dobosz K, Mikołajewski D (2013) Autism and ADHD–two ends of the same spectrum? Lect Notes Comput Sci 8226:623–630, 2013
10. Duch W, Dobosz K (2011) Visualization for understanding of neurodynamical systems. Cogn Neurodyn 5(2):145–160
11. Duch W, Nowak W, Meller J, Osiński G, Dobosz K, Mikołajewski D, Wójcik GM (2012) Computational approach to understanding autism spectrum disorders. Comput Sci 13:47–61
12. Duménieu M, Oulé M, Kreutz MR, Lopez-Rojas J (2017) The segregated expression of voltage-gated potassium and sodium channels in neuronal membranes: functional implications and regulatory mechanisms. Front Cell Neurosci 11:115
13. Elsabbagh M, Fernandes J, Jane Webb S, Dawson G, Charman T, Johnson MH (2013) Disengagement of visual attention in infancy is associated with emerging autism in toddlerhood. Biol Psychiatry 74(3):189–194
14. Gepner B, Féron F (2009) Autism: a world changing too fast for a miswired brain? Neurosci Biobehav Rev 33(8):1227–1242
15. Gravier A, Quek C, Duch W, Wahab A, Gravier-Rymaszewska J (2016) Neural network modelling of the influence of channelopathies on reflex visual attention. Cogn Neurodyn 10(1):49–72
16. Grossberg S, Seidman D (2006) Neural dynamics of autistic behaviors: cognitive, emotional, and timing substrates. Psychol Rev 113(3):483–525
17. Guglielmi L, Servettini I, Caramia M, Catacuzzeno L, Franciolini F, D'Adamo MC, Pessia M (2015) Update on the implication of potassium channels in autism: K+ channel autism spectrum disorder. Front Cell Neurosci 9:34
18. Heine M, Ciuraszkiewicz A, Voigt A, Heck J, Bikbaev A (2016) Surface dynamics of voltage-gated ion channels. Channels 10(4):267–281
19. Just MA, Keller TA, Malave VL, Kana RK, Varma S (2012) Autism as a neural systems disorder: a theory of frontal-posterior underconnectivity. Neurosci Biobehav Rev 36(4):1292–1313
20. Kawakubo Y, Kasai K, Okazaki S, Hosokawa-Kakurai M, Watanabe K, Kuwabara H, Ishijima M, Yamasue H, Iwanami A, Kato N, Maekawa H (2007) Electrophysiological abnormalities of spatial attention in adults with autism during the gap overlap task. Clin Neurophysiol 118:1464–1471
21. Kumar P, Kumar D, Jha SK, Jha NK, Ambasta RK (2016) Ion channels in neurological disorders. In: Donev R (ed) Advances in protein chemistry and structural biology, vol 103. Academic, Waltham, pp 97–136
22. Lai HC, Jan LY (2006) The distribution and targeting of neuronal voltage-gated ion channels. Nat Rev Neurosci 7(7):548–562
23. Landry R, Bryson SE (2004) Impaired disengagement of attention in young children with autism. J Child Psychol Psychiatry 45(6):1115–1122
24. O'Reilly RC, Hazy TE, Herd SA (2016) The Leabra cognitive architecture: how to play 20 principles with nature and win! In: Chipman S (ed) Oxford handbook of cognitive science. Oxford University Press, Oxford
25. O'Reilly RC, Munakata Y (2000) Computational explorations in cognitive neuroscience. MIT-Press, Cambridge
26. Rogers SJ, Ozonoff S (2005) Annotation: what do we know about sensory dysfunction in autism? A critical review of the empirical evidence. J Child Psychol Psychiatry 46(12):1255–1268
27. Thelen E, Smith LB (1996) A dynamic systems approach to the development of cognition and action. MIT Press, Cambridge
28. Uddin LQ, Supekar K, Menon V (2013) Reconceptualizing functional brain connectivity in autism from a developmental perspective. Front Hum Neurosci 7:458
29. Wen Y, Alshikho MJ, Herbert MR (2016) Pathway network analyses for autism reveal multisystem involvement, major overlaps with other diseases and convergence upon MAPK and calcium signaling. PLoS One 11(4):e0153329
30. Zimmerman AW (ed) (2008) Autism: current theories and evidence. Humana Press, Totowa

Part III
Memory Disorders

Chapter 14
Alzheimer's Disease: Rhythms, Local Circuits, and Model-Experiment Interactions

Frances K. Skinner and Alexandra Chatzikalymniou

Abstract As more biological details emerge from sophisticated experimental techniques today, we are faced with the increasing challenge of how best to develop and use computational models to gain insight into neurological diseases. In this chapter we briefly describe what is known regarding Alzheimer's disease (AD) and changes in brain rhythms as well as computational models in AD. We then briefly describe an expansion of our previous proposal of using whole hippocampus experimental preparations that spontaneously express θ and γ rhythms when developing microcircuit models. In this way, a cycling between model and experiment becomes possible allowing model insights to be brought to bear in understanding AD in our complex brain circuits.

14.1 Opening

The multi-scale, nonlinear, and detailed nature of human brain dynamics is what makes it complex and challenging to model and understand [1]. Also, as multi-scale interactions are thought to be a defining feature of brain functioning [2], they need to be explicitly considered. From a disease perspective, it is clear that cellular specifics require consideration [3, 4]. Moreover, we are now firmly in an age where inherent biological variability has been shown to be a part of the individuality of biological circuits and should be considered in trying to understand the boundaries between

F. K. Skinner (✉)
Krembil Research Institute, University Health Network, Toronto, ON, Canada

Department of Medicine (Neurology) and Physiology, University of Toronto, Toronto, ON, Canada
e-mail: frances.skinner@uhnresearch.ca; frances.skinner@utoronto.ca

A. Chatzikalymniou
Krembil Research Institute, University Health Network, Toronto, ON, Canada

Department of Physiology, University of Toronto, Toronto, ON, Canada
e-mail: alex4@windowslive.com

© Springer Nature Switzerland AG 2019
V. Cutsuridis (ed.), *Multiscale Models of Brain Disorders*, Springer Series in Cognitive and Neural Systems 13, https://doi.org/10.1007/978-3-030-18830-6_14

health and disease [5]. The need and advantage of computational models relative to experimental studies is their ability to examine and analyze output at multiple levels simultaneously so as to help understand the multi-scale, nonlinear underpinnings of neurological disease. This necessarily depends on the computational model's representation of brain circuits.

In this chapter, we focus on presenting Alzheimer's disease (AZ) from a *rhythmic* perspective and end with an update on a previous proposal [6].

14.2 Alzheimer's Disease and Rhythms

Alzheimer's disease (AD) is a devastating disorder, and early diagnosis is pivotal for its effective treatment. The community is in search of early biomarkers that can detect changes associated with early stages of the disease. AD seems to be a disorder of mechanisms underlying structural brain self-organization [7] and changes in brain connectivity [8]. Changes in subcortical structures in early- versus late-onset AD have also been shown [9] as well as early abnormalities in brain microstructure [10]. The main neuropathological model of the disease is the amyloid cascade model [11, 12] which in its final stages leads to neuronal death and decreases in brain volume [13]. Several amyloid-based AD models have been developed in rodents and non-rodents [14]. However, prior to amyloid segregation, changes in other kinds of brain activity occur and can serve as signals of the pathological changes.

Brain rhythms undergo fundamental changes in AD as seen in EEG recordings, in both rodents and humans. The hallmark of EEG abnormalities in AD patients is a shift of the power spectrum to lower frequencies [15–17] and a decrease in coherence of fast rhythms. Other than these general observations, changes occur in a frequency band-specific way. For example, in [18], delta activity was a significantly greater percentage of total EEG power in the moderate-to-advanced AD subjects when compared to either the healthy controls or mild AD subjects.

Besides changes in spectral EEG power measurements in AD, changes in additional measurements of neural activity have been observed in AD. Phase amplitude coupling (PAC) between a slower and a faster rhythm, a form of cross frequency coupling, is one of these additional measurements. PAC is associated with memory performance and changes in AD [19]. In [20], PAC between alpha and gamma was altered in AD patients, and in [21], altered θ-γ PAC was found in individuals with AD and mild cognitive impairment (MCI). Global field synchronization (GFS) is a commonly used measure of EEG synchronization and reflects the global amount of phase-locked activity at a given frequency. In [22], patients showed decreased GFS values in α, β, and γ frequency bands and increased GFS values in the delta band, supporting a hypothesized functional disconnection in neurocognitive networks. In another study, GFS values were found to be significantly lower in AD patients relative to healthy controls [23].

In essence, EEG abnormalities of AD patients are characterized by slowed mean frequency and PAC changes, less complex activities, and reduced coherence among cortical regions [24]. Overall, the changes in EEG suggest that it has utility as a valuable tool for early diagnosis of AD and could be especially helpful in combination with other biological/neuropsychological markers and structural/functional imaging [15, 25–28]. Particular rhythmic changes occur in local field potential (LFP) recordings of AD subjects [29]. Interestingly, changes in PAC of θ and γ LFP rhythms in the hippocampus have been found to precede pathological changes, thus indicating that it could serve as an early biomarker [30–33]. These changes have been shown to be specific to different brain structure regions that include the hippocampus and parietal cortex, but not the prefrontal cortex [34].

14.2.1 Lower Level Changes

Synaptic malfunctions and ion channel expression changes are observed in AD. Synapse loss is an early and invariant feature of AD, and there is a strong correlation between the extent of synapse loss and the severity of dementia [35]. Rubio et al. [36] showed that older (8 months) behaving mice expressing mutated human amyloid precursor protein (hAPP) had diminished θ and γ rhythm power and a significant deficit in GABAergic septo-hippocampal (SH) innervation as compared to aged normal mice, in addition to the well-known loss of cholinergic input to the hippocampus in AD. This was shown to be due to a reduced number and complexity of SH axons and not neuronal loss.

Verret et al. [37] showed that reduced Nav1.1 levels and parvalbumin cell dysfunction critically contributed to abnormalities in oscillatory rhythms, network synchrony, and memory in hAPP mice and possibly in AD. Moreover, restoring Nav1.1 levels in hAPP mice by Nav1.1 expression increased inhibitory synaptic activity and γ oscillations and reduced hypersynchrony, memory deficits, and premature mortality. Restoration of brain rhythms and cognition has also been shown in Nav1.1-overexpressing interneuron transplants in mouse AD models [38].

14.2.2 Computational Models and AD

The challenge of multi-scale modeling aspects in neuroscience is exposed by the various modeling studies that have been done in AD [39]. Subcellular, cellular, circuit, and system-level models in consideration of AD have been built. Neural mass models were used to be able to consider whole brain representations and to be able to examine the hypothesis that excessive neuronal activity leads to degeneration and hub vulnerability in AD [40]. At a different level, cross talk between multiple cell types including neurons and glia and amyloid-beta (Aβ) was modeled to help understand the neurodegenerative progression of AD [41].

A modeling focus has been on the hippocampus. For example, Menschik and Finkel [42] built circuit models of the CA3 region of the hippocampus in consideration of memory alterations, and Cutsuridis et al. [43] considered encoding and retrieval aspects in the CA1 region that could be used for AD insights. The specific effect of Aβ on synaptic release probability was examined in CA1 pyramidal cell models [44], and a hippocampo-septal circuit model was used by [45] to examine changes in θ band power when Aβ changes in pyramidal cell ion channels were modeled.

14.3 Proposal Redux

As described above, rhythmic changes as captured in EEG and LFP measures in AD are clear, as well as cellular and synaptic specific changes. Models that examine AD aspects have been developed and some examples are provided above. Model considerations range from biochemical signaling to system-level processing [39].

How might one further bring forth the advantage of computational modeling to help understand the multi-scale, nonlinear complexities that underlie these multiple level changes? Incorporating particular biological details (e.g., particular ion channel biophysics in a given cell type) typically leads to one having sparse experimental data (e.g., other ion channel types and their biophysical characteristics in the given cell type as well as ion channel biophysics for other cell types) in building microcircuit models. As such, a focus on whole brain EEG modeling with a cellular representation is not sensible. Further, a focus on LFP rather than EEG recordings in models would help reduce the spatial extent and complexity [46]. Given the present age in which we exist with updated knowledge and ever-expanding information about different cell types and the availability of molecular datasets and more [47], we think that the most critical aspect is to be able to have a cycling between modeling work and experimental data so that additional knowledge can be continually considered and possibly incorporated to obtain additional insights from models.

We previously suggested that a focus on microcircuit modeling where direct links between model and experiment at multiple levels are possible can better leverage insights gained from computational modeling [6]. This is possible if a model focus on an *in vitro* whole hippocampus preparation is used in which spontaneous θ and θ/γ rhythms are expressed [48, 49]. That these rhythms emerge spontaneously suggests that they form part of the natural output of the biological system. We are using such a model focus and now have a working mechanism for the generation of θ rhythms in this preparation that includes fast-firing parvalbumin interneurons and pyramidal cells with spike-frequency adaptation and post-inhibitory rebound characteristics in CA1 microcircuits [50]. We have also built models to examine the contribution of additional inhibitory cell types to *ongoing* theta rhythms [51]. We have also leveraged these latter models to build biophysical LFP models that were constrained with LFP experimental characteristics leading to insights of how different cell types and pathways could contribute to robust LFP theta rhythms in CA1 microcircuits [52]. This latter work is a critical development and is schematized in Fig. 14.1 that we consider as an updated proposal from [6]. That is,

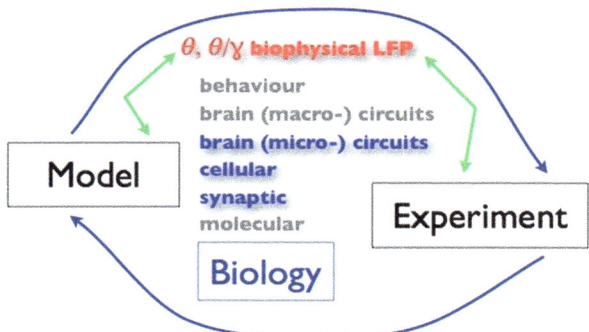

Fig. 14.1 Proposal Redux Schematic. Biophysical LFP models and experiments (green arrows) of rhythmic activities (θ, θ/γ) from microcircuits in which the models have biological representations that span more than one scale (middle blue wordings) and in which there are direct linkages between model and experiment. A cycling (blue arrows) between model and experiment is then possible. Based on [6]

a cycling between model and experiment (blue arrows) in which there are biological representations (blue wordings) in the model microcircuits of synaptic, cellular, and network levels can occur. The cycling is possible because comparisons between model and experiment are possible at multiple levels, importantly including the LFP output from the microcircuit (green arrows) of the isolated whole hippocampus preparation. Therefore, any model predictions or developed hypotheses can be examined in the experimental system.

Moving forward, the CA1 hippocampus microcircuit model can be expanded to include additional levels (e.g., molecular, as shown in gray wording in Fig. 14.1) incorporating further data and considering other detailed models [53] but always being able to be directly constrained by the experimental system of the isolated whole hippocampus preparation. As this preparation has been specifically used to show particular θ and θ/γ changes using mouse AD models [30], any insights obtained by our models have the possibility of providing insight into the complex circuit dynamics of AD states.

References

1. Marder E (2015) Understanding brains: details, intuition, and big data.   PLoS Biol 13(5):e1002147. https://doi.org/10.1371/journal.pbio.1002147
2. Cohen MX, Gulbinaite R (2014) Five methodological challenges in cognitive electrophysiology. NeuroImage 85(Part 2):702–710. https://doi.org/10.1016/j.neuroimage.2013.08.010
3. Ferguson BR, Gao WJ (2018) PV interneurons: critical regulators of E/I balance for prefrontal cortex-dependent behavior and psychiatric disorders. Front Neural Circuits 12. https://doi.org/10.3389/fncir.2018.00037

4. Palop JJ, Mucke L (2016) Network abnormalities and interneuron dysfunction in Alzheimer disease. Nat Rev Neurosci 17(12):777–792. https://doi.org/10.1038/nrn.2016.141
5. Grashow R, Brookings T, Marder E (2009) Reliable neuromodulation from circuits with variable underlying structure. Proc Natl Acad Sci U S A 106(28). https://doi.org/10.1073/pnas.0905614106
6. Skinner FK, Ferguson KA (2013) Modeling oscillatory dynamics in brain microcircuits as a way to help uncover neurological disease mechanisms: a proposal. Chaos: Interdiscip J Nonlinear Sci 23(4):046108. https://doi.org/10.1063/1.4829620
7. Arendt T (2001) Alzheimer's disease as a disorder of mechanisms underlying structural brain self-organization. Neuroscience 102(4):723–765
8. Supekar K, Menon V, Rubin D, Musen M, Greicius MD (2008) Network analysis of intrinsic functional brain connectivity in Alzheimer's disease. PLoS Comput Biol 4(6). https://doi.org/10.1371/journal.pcbi.1000100
9. Cho H, Seo SW, Kim JH, Kim C, Ye BS, Kim GH, Noh Y, Kim HJ, Yoon CW, Seong JK, Kim CH, Kang SJ, Chin J, Kim ST, Lee KH, Na DL (2013) Changes in subcortical structures in early- versus late-onset Alzheimer's disease. Neurobiol Aging 34(7):1740–1747. https://doi.org/10.1016/j.neurobiolaging.2013.01.001
10. Douaud G, Menke RAL, Gass A, Monsch AU, Rao A, Whitcher B, Zamboni G, Matthews PM, Sollberger M, Smith S (2013) Brain microstructure reveals early abnormalities more than two years prior to clinical progression from mild cognitive impairment to Alzheimer's disease. J Neurosci 33(5):2147–2155. https://doi.org/10.1523/JNEUROSCI.4437-12.2013
11. Boche D, Nicoll JAR (2010) Are we getting to grips with Alzheimer's disease at last? Brain J Neurol 133(Pt 5):1297–1299. https://doi.org/10.1093/brain/awq099
12. Minati L, Edginton T, Bruzzone MG, Giaccone G (2009) Current concepts in Alzheimer's disease: a multidisciplinary review. Am J Alzheimer's Dis Other Dement 24(2):95–121. https://doi.org/10.1177/1533317508328602
13. O'Dwyer L, Lamberton F, Matura S, Tanner C, Scheibe M, Miller J, Rujescu D, Prvulovic D, Hampel H (2012) Reduced hippocampal volume in healthy young ApoE4 carriers: an MRI study. PloS One 7(11):e48895. https://doi.org/10.1371/journal.pone.0048895
14. Saraceno C, Musardo S, Marcello E, Pelucchi S, Diluca M (2013) Modeling Alzheimer's disease: from past to future. Exp Pharmacol Drug Discov 4:77. https://doi.org/10.3389/fphar.2013.00077
15. Besthorn C, Zerfass R, Geiger-Kabisch C, Sattel H, Daniel S, Schreiter-Gasser U, Förstl H (1997) Discrimination of Alzheimer's disease and normal aging by EEG data. Electroencephalogr Clin Neurophysiol 103(2):241–248
16. Coben LA, Danziger WL, Berg L (1983) Frequency analysis of the resting awake EEG in mild senile dementia of Alzheimer type. Electroencephalogr Clin Neurophysiol 55(4):372–380
17. Moretti DV, Babiloni C, Binetti G, Cassetta E, Dal Forno G, Ferreric F, Ferri R, Lanuzza B, Miniussi C, Nobili F, Rodriguez G, Salinari S, Rossini PM (2004) Individual analysis of EEG frequency and band power in mild Alzheimer's disease. Clin Neurophysiol 115(2):299–308
18. Hier DB, Mangone CA, Ganellen R, Warach JD, Van Egeren R, Perlik SJ, Gorelick PB (1991) Quantitative measurement of delta activity in Alzheimer's disease. Clin EEG (electroencephalogr) 22(3):178–182
19. Bergmann TO, Born J (2018) Phase-amplitude coupling: a general mechanism for memory processing and synaptic plasticity? Neuron 97(1):10–13. https://doi.org/10.1016/j.neuron.2017.12.023
20. Poza J, Bachiller A, Gomez C, Garcia M, Nunez P, Gomez-Pilar J, Tola-Arribas MA, Cano M, Hornero R (2017) Phase-amplitude coupling analysis of spontaneous EEG activity in Alzheimer's disease. In: Conference proceedings: ...annual international conference of the IEEE engineering in medicine and biology society. IEEE Engineering in Medicine and Biology Society, pp 2259–2262. https://doi.org/10.1109/EMBC.2017.8037305
21. Goodman MS, Kumar S, Zomorrodi R, Ghazala Z, Cheam ASM, Barr MS, Daskalakis ZJ, Blumberger DM, Fischer C, Flint A, Mah L, Herrmann N, Bowie CR, Mulsant BH, Rajji TK (2018) Theta-gamma coupling and working memory in Alzheimer's dementia and mild cognitive impairment. Frontiers Aging Neurosci 10. https://doi.org/10.3389/fnagi.2018.00101

22. Koenig T, Prichep L, Dierks T, Hubl D, Wahlund LO, John ER, Jelic V (2005) Decreased EEG synchronization in Alzheimer's disease and mild cognitive impairment. Neurobiol Aging 26(2):165–171. https://doi.org/10.1016/j.neurobiolaging.2004.03.008

23. Park YM, Che HJ, Im CH, Jung HT, Bae SM, Lee SH (2008) Decreased EEG synchronization and its correlation with symptom severity in Alzheimer's disease. Neurosci Res 62(2):112–117. https://doi.org/10.1016/j.neures.2008.06.009

24. Wang J, Fang Y, Wang X, Yang H, Yu X, Wang H (2017) Enhanced gamma activity and cross-frequency interaction of resting-state electroencephalographic oscillations in patients with Alzheimer's disease. Front Aging Neurosci 9. https://doi.org/10.3389/fnagi.2017.00243

25. Alberdi A, Aztiria A, Basarab A (2016) On the early diagnosis of Alzheimer's disease from multimodal signals: a survey. Artif Intell Med 71:1–29. https://doi.org/10.1016/j.artmed.2016.06.003

26. Dauwels J, Vialatte F, Cichocki A (2010) Diagnosis of Alzheimer's disease from EEG signals: where are we standing? Curr Alzheimer Res 7(6):487–505

27. Jeong J (2004) EEG dynamics in patients with Alzheimer's disease. Clin Neurophysiol 115(7):1490–1505. https://doi.org/10.1016/j.clinph.2004.01.001

28. Lizio R, Vecchio F, Frisoni GB, Ferri R, Rodriguez G, Babiloni C (2011) Electroencephalographic rhythms in Alzheimer's disease. Int J Alzheimer's Dis 2011:927573. https://doi.org/10.4061/2011/927573

29. Kitchigina VF (2018) Alterations of coherent theta and gamma network oscillations as an early biomarker of temporal lobe epilepsy and Alzheimer's disease. Front Integr Neurosci 12. https://doi.org/10.3389/fnint.2018.00036

30. Goutagny R, Gu N, Cavanagh C, Jackson J, Chabot JG, Quirion R, Krantic S, Williams S (2013) Alterations in hippocampal network oscillations and theta-gamma coupling arise before Aβ overproduction in a mouse model of Alzheimer's disease. Eur J Neurosci 37(12):1896–1902. https://doi.org/10.1111/ejn.12233

31. Goutagny R, Krantic S (2013) Hippocampal oscillatory activity in Alzheimer's disease: toward the identification of early biomarkers? Aging Dis 4(3):134–140

32. Hamm V, Héraud C, Cassel JC, Mathis C, Goutagny R (2015) Precocious alterations of brain oscillatory activity in Alzheimer's disease: a window of opportunity for early diagnosis and treatment. Front Cell Neurosci 491. https://doi.org/10.3389/fncel.2015.00491

33. Mondragón-Rodríguez S, Gu N, Manseau F, Williams S (2018) Alzheimer's transgenic model is characterized by very early brain network alterations and -CTF fragment accumulation: reversal by -secretase inhibition. Front Cell Neurosci 12. https://doi.org/10.3389/fncel.2018.00121

34. Zhang X, Zhong W, Brankačk J, Weyer SW, Müller UC, Tort ABL, Draguhn A (2016) Impaired theta-gamma coupling in APP-deficient mice. Sci Rep 6:21948. https://doi.org/10.1038/srep21948

35. Shankar GM, Walsh DM (2009) Alzheimer's disease: synaptic dysfunction and Abeta. Mol Neurodegener 4:48. https://doi.org/10.1186/1750-1326-4-48

36. Rubio SE, Vega-Flores G, Martínez A, Bosch C, Pérez-Mediavilla A, del Río J, Gruart A, Delgado-García JM, Soriano E, Pascual M (2012) Accelerated aging of the GABAergic septo-hippocampal pathway and decreased hippocampal rhythms in a mouse model of Alzheimer's disease. FASEB J 26(11):4458–4467. https://doi.org/10.1096/fj.12-208413

37. Verret L, Mann EO, Hang GB, Barth AMI, Cobos I, Ho K, Devidze N, Masliah E, Kreitzer AC, Mody I, Mucke L, Palop JJ (2012) Inhibitory interneuron deficit links altered network activity and cognitive dysfunction in Alzheimer model. Cell 149(3):708–721. https://doi.org/10.1016/j.cell.2012.02.046

38. Martinez-Losa M, Tracy TE, Ma K, Verret L, Clemente-Perez A, Khan AS, Cobos I, Ho K, Gan L, Mucke L, Alvarez-Dolado M, Palop JJ (2018) Nav1.1-overexpressing interneuron transplants restore brain rhythms and cognition in a mouse model of Alzheimer's disease. Neuron 98(1):75–89.e5. https://doi.org/10.1016/j.neuron.2018.02.029

39. Cutsuridis V, Moustafa AA (2017) Computational models of Alzheimer's disease. Scholarpedia 12(1):32144. https://doi.org/10.4249/scholarpedia.32144

40. de Haan W, Mott K, van Straaten ECW, Scheltens P, Stam CJ (2012) Activity dependent degeneration explains hub vulnerability in Alzheimer's disease. PLoS Comput Biol 8(8):e1002582. https://doi.org/10.1371/journal.pcbi.1002582
41. Puri IK, Li L (2010) Mathematical modeling for the pathogenesis of Alzheimer's disease. PLOS ONE 5(12):e15176. https://doi.org/10.1371/journal.pone.0015176
42. Menschik ED, Finkel LH (1998) Neuromodulatory control of hippocampal function: towards a model of Alzheimer's disease. Artif Intell Med 13(1–2):99–121
43. Cutsuridis V, Cobb S, Graham BP (2010) Encoding and retrieval in a model of the hippocampal CA1 microcircuit. Hippocampus 20(3):423–446. https://doi.org/10.1002/hipo.20661
44. Romani A, Marchetti C, Bianchi D, Leinekugel X, Poirazi P, Migliore M, Marie H (2013) Computational modeling of the effects of amyloid-beta on release probability at hippocampal synapses. Front Comput Neurosci 7:1. https://doi.org/10.3389/fncom.2013.00001
45. Zou X, Coyle D, Wong-Lin K, Maguire L (2011) Computational study of hippocampal-septal theta rhythm changes due to beta-amyloid-altered ionic channels. PLoS ONE 6(6):e21579. https://doi.org/10.1371/journal.pone.0021579
46. Cohen MX (2017) Where does EEG come from and what does it mean? Trends Neurosci 40(4):208–218. https://doi.org/10.1016/j.tins.2017.02.004
47. Koroshetz W, Gordon J, Adams A, Beckel-Mitchener A, Churchill J, Farber G, Freund M, Gnadt J, Hsu NS, Langhals N, Lisanby S, Liu G, Peng GCY, Ramos K, Steinmetz M, Talley E, White S (2018) The state of the NIH BRAIN initiative. J Neurosci 38(29):6427–6438. https://doi.org/10.1523/JNEUROSCI.3174-17.2018
48. Goutagny R, Jackson J, Williams S (2009) Self-generated theta oscillations in the hippocampus. Nat Neurosci 12(12):1491–1493. https://doi.org/10.1038/nn.2440
49. Jackson J, Goutagny R, Williams S (2011) Fast and slow gamma rhythms are intrinsically and independently generated in the subiculum. J Neurosci 31(34):12104–12117. https://doi.org/10.1523/JNEUROSCI.1370-11.2011
50. Ferguson KA, Chatzikalymniou AP, Skinner FK (2017) Combining theory, model, and experiment to explain how intrinsic theta rhythms are generated in an in vitro whole hippocampus preparation without oscillatory inputs. eNeuro 4(4). https://doi.org/10.1523/ENEURO.0131-17.2017
51. Ferguson KA, Huh CYL, Amilhon B, Manseau F, Williams S, Skinner FK (2015) Network models provide insights into how oriens-lacunosum-moleculare and bistratified cell interactions influence the power of local hippocampal CA1 theta oscillations. Frontiers Syst Neurosci 9:110. https://doi.org/10.3389/fnsys.2015.00110
52. Chatzikalymniou AP, Skinner FK (2018) Deciphering the contribution of Oriens-Lacunosum/Moleculare (OLM) cells to intrinsic theta rhythms using biophysical local field potential (LFP) models. eNeuro pp ENEURO.0146–18.2018. https://doi.org/10.1523/ENEURO.0146-18.2018
53. Bezaire MJ, Raikov I, Burk K, Vyas D, Soltesz I (2016) Interneuronal mechanisms of hippocampal theta oscillation in a full-scale model of the rodent CA1 circuit. eLife 5:e18566. https://doi.org/10.7554/eLife.18566

Chapter 15
Using a Neurocomputational Autobiographical Memory Model to Study Memory Loss

Di Wang, Ahmed A. Moustafa, Ah-Hwee Tan, and Chunyan Miao

Abstract Autobiographical memory (AM) is a core component of human life and plays an important role in self-identification. Various conceptual models have been proposed to explain its functionalities and describe its dynamics. However, most existing computational AM models do not distinguish AM from other long-term memory. Specifically, during model design, the unique features and the memory encoding, storage, and retrieval procedures of AM were not taken into consideration in prior models. In this chapter, we introduce our neurocomputational AM model, which is consistent with Conway and Pleydell-Pearce's model in terms of both the network structure and dynamics. We further propose how to apply our parameterized computational model to quantitatively study memory loss in people of different age groups. As such, we provide a suitable tool to evaluate the effect of different memory loss phases in a rapid and quantitative manner, which may be difficult or impossible in experimental studies on human subjects.

Keywords Autobiographical memory · Neurocomputational model · Memory loss · Cognitive modeling

D. Wang (✉)
Joint NTU-UBC Research Centre of Excellence in Active Living for the Elderly, Nanyang Technological University, Singapore
e-mail: wangdi@ntu.edu.sg

A. A. Moustafa
School of Social Sciences and Psychology & Marcs Institute for Brain and Behaviour, Western Sydney University, Sydney, New South Wales, Australia
e-mail: a.moustafa@westernsydney.edu.au

A.-H. Tan · C. Miao
Joint NTU-UBC Research Centre of Excellence in Active Living for the Elderly, Nanyang Technological University, Singapore

School of Computer Science and Engineering, Nanyang Technological University, Singapore
e-mail: asahtan@ntu.edu.sg; ascymiao@ntu.edu.sg

© Springer Nature Switzerland AG 2019
V. Cutsuridis (ed.), *Multiscale Models of Brain Disorders*, Springer Series in Cognitive and Neural Systems 13, https://doi.org/10.1007/978-3-030-18830-6_15

15.1 Benefits of Modeling AM Using a Neurocomputational Model

Autobiographical memory (AM) is "a system that encodes, stores and guides retrieval of all episodic information related to our personal experiences" [4]. It is a core component of human brain function and plays an important role in self-identification. Specifically, "individuals' current self-views, beliefs, and goals influence their collections and appraisals of former selves. In turn, people's current self-views are influenced by what they remember about their personal past, as well as how they recall earlier selves and episodes" [32]. Due to its great importance, over decades, researchers from different disciplines have tried to find out the working mechanisms of AM. Although until today, we still do not fully understand the dynamics of AM on the neural network level, we have already learned about its activation regions in the brain using various neural imaging techniques [7]. Moreover, some conceptual AM models proposed in the literature have been supported by neural imaging evidence (e.g., Conway and Pleydell-Pearce's model [8] has been supported by [1]).

Among the various AM models established by psychologists, the one proposed by Conway and Pleydell-Pearce [8] has been widely accepted in the academic world. They categorized autobiographical memory knowledge into three levels, namely, *lifetime periods, general events*, and *event-specific knowledge* (from general to specific, see Fig. 15.1). Furthermore, they proposed that autobiographical memories can be directly accessed if the cues are specific and personally relevant. On the

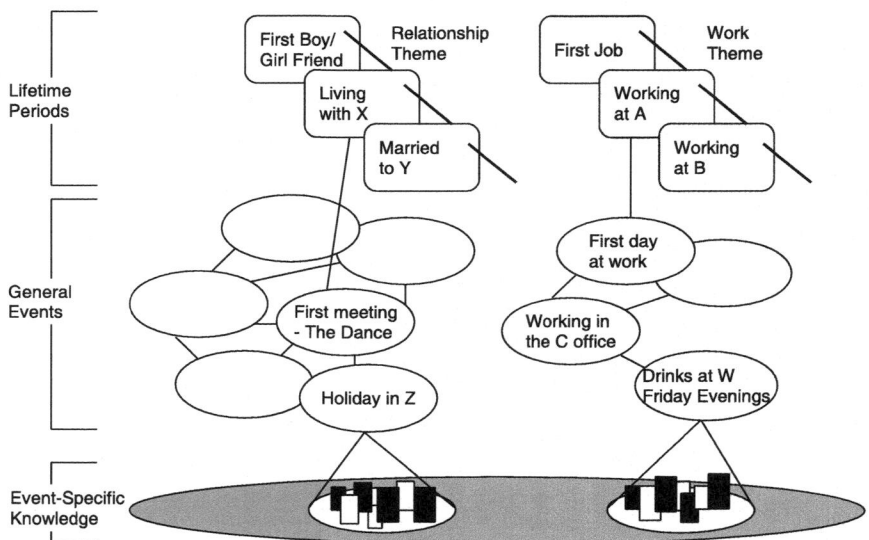

Fig. 15.1 Illustration of the autobiographical memory hierarchy proposed by [8]. (This figure replicates Figure 1 presented in [8])

other hand, if the cues are general, a *generative retrieval* process must be engaged to produce more specific cues for the retrieval of relevant memories. The main difference between direct and generative retrievals is that "the search process is modulated by control processes in generative retrieval but not, or not so extensively, in direct retrieval" [8]. We adopt these established theories as the design principles of our computational AM model, which is introduced in the following section.

Face-to-face interview is the conventional way of finding out how much AM a human subject can recall or which type of AM does one recall easily. Inevitably, a large amount of effort is involved in this process. For example, the study reported in [3] recruited 130 interviewers to record the response of 1,241 participants, and the study conducted later [22] recruited 120 interviewers and 1,307 participants. Nonetheless, these studies only evaluate our overall capability of memory recalls across different life stages, which do not distinguish the respective effect of the memory encoding, storage, and retrieval phases. On the other hand, advanced neural imaging techniques allow us to learn the activation regions and the sequence of activations in our brain. However, these studies only reveal the dynamics of memory retrieval and may not be suitable to study memory encoding and storage. If we can build a neurocomputational model based on prior neurocognitive studies, we can then quantitatively evaluate various phenomena related to AM in a rapid and quantitative manner, which may be difficult or impossible in human participants. For example, by using a parameterized computational model, we may not need to conduct thorough interviews to learn one's memory performance. Instead, with the help of the peripheral information such as age, gender, education level, medical history, etc., we may only need to ask few questions to identify human subjects' key parameters to approximate their memory performance. Moreover, with a computational model, we can better design and deliver personalized reminiscence therapies [28, 31] to improve the psychological and cognitive well-being of the elderly [17]. In the following section, we introduce our neurocomputational AM model, which serves as an appropriate groundwork for future rapid, quantitative, and more complex evaluations on autobiographical memory loss.

15.2 Autobiographical Memory-Adaptive Resonance Theory (AM-ART) Network

Existing neurocomputational AM models (e.g., [6, 13, 20, 26]) may not be suitable for quantitative evaluations of memory loss because their memory retrieval mechanisms are over-simplified that they either simply retrieve all memories or retrieve all memories containing certain keyword(s). Other computational long-term memory models embedded in cognitive systems (e.g., [2, 15, 16]) may also not be suitable because they do not consider the emotional aspects [27] of the stored memory, which is one of the major differences between AM and other types of long-term memories. To enable quantitative evaluations of memory loss in a rapid manner, we introduce

Autobiographical Memory-Adaptive Resonance Theory (AM-ART) network [29], which is a formally defined neurocomputational model whose structure is consistent with Conway and Pleydell-Pearce's model (see Fig. 15.1). Interested readers please refer to [29] for all the mathematical formulations and algorithms.

The network structure of AM-ART is shown in Fig. 15.2. AM-ART is a three-layer neural network that in the top-down order, its F_3, F_2, and F_1 layers encode *lifetime periods, general events*, and *event-specific knowledge*, respectively. AM-ART is designed based on Conway and Pleydell-Pearce's model that we can highlight the correspondence between the two models using the following examples. The life experience of "working at A" can be represented as a code (learned episode) in F_3 of AM-ART. The associated events of that episode, namely, "first day at work," "working in the C office," and "drinks at W Friday evenings," can be represented as codes (learned events) in F_2. A specific event, taking "drinks at W Friday evenings" as an example, can be read out in F_1 that on Friday night (time), at W (location), with colleagues (people), drinking (activity), feeling happy (emotion), together with the pictorial memory (imagery). Furthermore, memory retrieval in AM-ART replicates the three stages of the generative memory retrieval presented in [8], namely, the *elaboration stage, strategic search stage*, and *evaluation stage*. The operations applied in AM-ART to realize the three stages of the generative memory retrieval are summarized in Table 15.1.

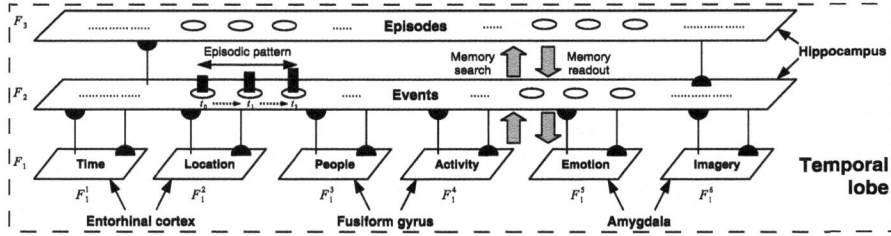

Fig. 15.2 Network structure of AM-ART. All its channels and layers match specific brain regions

Table 15.1 Operations in AM-ART to realize the generative autobiographical memory retrieval

#	Stage	Description in [8]	AM-ART operations in [29]
1	Elaboration	"The elaboration of a cue with which to search memory and the simultaneous setting of verification criteria."	Template masking, mutation, and setting of the vigilance parameters
2	Strategic search	"Matching the description to records in memory."	Code activation and code competition
3	Evaluation	"Records accessed in memory were assessed against the verification criteria."	Template matching

Moreover, in terms of functional mapping in human brain, we find that AM-ART would reside in the temporal lobe (see Fig. 15.2). Specifically, inputs of time and location are from entorhinal cortex [14], inputs of people and activity are from fusiform gyrus [12], inputs of emotion and imagery are from amygdala [18], and both the F_2 and F_3 layers reside in hippocampus [24]. Note that the inputs to AM-ART are considered as recognized or processed information, e.g., imagery used for memory encoding in hippocampus is from amygdala [18] rather than directly from occipital lobe.

15.3 Using AM-ART to Study Autobiographical Memory Loss

Memory loss generally occurs during three phases, namely, memory formation, storage, and retrieval [11]. Specifically, we can introduce three parameters to AM-ART to regulate the corresponding memory loss processes, namely, *overload* as the intensity of demanding cognitive tasks during formation [9], *decay* as the rate of long-term memory fading during storage [21], and *inhibition* as the likelihood of retrieval failure during retrieval [25]. This way of using parameters to model memory loss is in line with cognitive studies that "the individual pattern of impaired memory functions correlates with parameters of structural or functional brain integrity" [11].

After the identification of the three memory loss parameters, the next step is to estimate their values among different types of people based on prior surveys, such as the results reported in [3, 5, 10, 19, 22, 23]. These survey data may be used interchangeably for training, i.e., to estimate the memory loss parameter values among different age groups, and testing, i.e., to evaluate whether the estimated parameter values can be used for memory loss predictions. Note that the values of the memory loss parameters may influence the overall performance in a nonlinear manner. Interested readers please refer to [30] for the preliminary results.

The proposed overload, decay, and inhibition parameters may only represent one possible combination of the underlying mechanisms of memory loss. Based on AM-ART, other memory loss mechanisms may also be tested. For example, some researchers hypothesize that the formed long-term memories will not biologically get lost. However, we simply cannot retrieve them back. This possibility can be easily emulated using AM-ART by excluding memory losses during storage and making it harder when retrieve the old memories.

Acknowledgements This research is supported, in part, by the National Research Foundation, Prime Minister's Office, Singapore, under its IDM Futures Funding Initiative and the Singapore Ministry of Health under its National Innovation Challenge on Active and Confident Ageing (NIC Project No. MOH/NIC/COG04/2017).

References

1. Addis DR, Knapp K, Roberts RP, Schacter DL (2012) Routes to the past: neural substrates of direct and generative autobiographical memory retrieval. Neuroimage 59(3):2908–2922
2. Anderson JR, Bothell D, Byrne MD, Douglass S, Lebiere C, Qin Y (2004) An integrated theory of the mind. Psychol Rev 111:1036–1060
3. Berntsen D, Rubin DC (2002) Emotionally charged autobiographical memories across the life span: the recall of happy, sad, traumatic, and involuntary memories. Psychol Aging 17(4):636–652
4. Bluck S, Levine LJ (1998) Reminiscence as autobiographical memory: a catalyst for reminiscence theory development. Ageing Soc 18(2):185–208
5. Bluck S, Levine LJ, Laulhere TM (1999) Autobiographical remembering and hypermnesia: a comparison of older and younger adults. Psychol Aging 14(4):671–682
6. Boloni L (2014) Integrating perception, narrative, premonition and confabulatory continuation. Biol Inspired Cogn Archit 8:120–131
7. Cabeza R, St Jacques P (2007) Functional neuroimaging of autobiographical memory. Trends Cogn Sci 11(5):219–227
8. Conway MA, Pleydell-Pearce CW (2000) The construction of autobiographical memories in the self-memory system. Psychol Rev 107(2):261–288
9. Daselaar SM, Prince SE, Dennis NA, Hayes SM, Kim H, Cabeza R (2009) Posterior midline and ventral parietal activity is associated with retrieval success and encoding failure. Front Hum Neurosci 3:1–10
10. Demiray B, Gulgoz S, Bluck S (2009) Examining the life story account of the reminiscence bump: why we remember more from young adulthood. Memory 17(7):708–723
11. Jahn H (2013) Memory loss in Alzheimer's disease. Dialogues Clin Neurosci 15:445–454
12. Kanwisher N (2001) Neural events and perceptual awareness. Cognition 79:89–113
13. Kope A, Rose C, Katchabaw M (2013) Modeling autobiographical memory for believable agents. In: Proceedings of AAAI conference on artificial intelligence and interactive digital entertainment, pp 23–29
14. Kraus BJ, Brandon MP, Robinson RJ II, Connerney MA, Hasselmo ME, Eichenbaum H (2015) During running in place, grid cells integrate elapsed time and distance run. Neuron 88(3):578–589
15. Laird JE (2012) The soar cognitive architecture. MIT Press, Cambridge/London
16. Langley P (2006) Cognitive architectures and general intelligent systems. AI Magazine 27:33–44
17. Lin YC, Dai YT, Hwang SL (2003) The effect of reminiscence on the elderly population: a systematic review. Public Health Nurs 20(4):297–306
18. Phelps EA (2004) Human emotion and memory: interactions of the amygdala and hippocampal complex. Curr Opin Neurobiol 14:198–202
19. Piolino P, Desgranges B, Clarys D, Guillery-Girard B, Taconnat L, Isingrini M, Eustache F (2006) Autobiographical memory, autonoetic consciousness, and self-perspective in aging. Psychol Aging 21(3):510–525
20. Pointeau G, Petit M, Dominey PF (2014) Successive developmental levels of autobiographical memory for learning through social interaction. IEEE Trans Auton Ment Dev 6(3):200–212
21. Rubin DC (1982) On the retention function for autobiographical memory. J Verbal Learn Verbal Behav 21(1):21–38
22. Rubin DC, Berntsen D (2003) Life scripts help to maintain autobiographical memories of highly positive, but not highly negative, events. Mem Cogn 31(1):1–14
23. Rubin DC, Schulkind MD (1997) Distribution of important and word-cued autobiographical memories in 20-, 35-, and 70-year-old adults. Psychol Aging 12(3):524–535
24. Stark SM, Yassa MA, Lacy JW, Stark CEL (2013) A task to assess behavioral pattern separation (BPS) in humans: data from healthy aging and mild cognitive impairment. Neuropsychologia 51(12):2442–2449

25. Storm BC, Levy BJ (2012) A progress report on the inhibitory account of retrieval-induced forgetting. Mem Cogn 40(6):827–843
26. Thudt A, Baur D, Huron S (2016) Visual mementos: reflecting memories with personal data. IEEE Trans Vis Comput Graph 22(1):369–378
27. Wang D, Tan AH (2014) Mobile humanoid agent with mood awareness for elderly care. In: Proceedings of international joint conference on neural networks, pp 1549–1556
28. Wang D, Tan AH (2015) MyLife: an online personal memory album. In: Proceedings of IEEE/WIC/ACM international conference on web intelligence and intelligent agent technology, pp 243–244
29. Wang D, Tan AH, Miao C (2016) Modelling autobiographical memory in human-like autonomous agents. In: Proceedings of international conference on autonomous agents and multiagent systems, pp 845–853
30. Wang D, Tan AH, Miao C, Moustafa AA (2019) Modelling autobiographical memory loss across life span. In: Proceedings of AAAI conference on artificial intelligence, to appear
31. Watt LM, Cappeliez P (2000) Integrative and instrumental reminiscence therapies for depression in older adults: intervention strategies and treatment effectiveness. Aging Ment Health 4(2):166–177
32. Wilson AE, Ross M (2003) The identity function of autobiographical memory: time is on our side. Memory 11(2):137–149

Part IV
Other Disorders

Chapter 16
How Can Computer Modelling Help in Understanding the Dynamics of Absence Epilepsy?

Piotr Suffczynski, Stiliyan Kalitzin, and Fernando H. Lopes da Silva

What I cannot create, I do not understand

Richard Feynman

Abstract An overview of the pathophysiology of absence seizures is given, focusing on computational modelling where recent neurophysiological experimental evidence is incorporated. The main question addressed is what is the dynamical process by which the same brain can produce sustained bursts of synchronous spike-and-wave discharges (SWDs) and normal, largely desynchronized brain activity, i.e. to display *bistability*. This generic concept, tested on an updated neural mass computational model of absence seizures, predicts certain properties of the probability distributions of inter-ictal intervals and of the durations of ictal events. A critical analysis of the distributions predicted by the model and those found in reality led to adjustments of the model with respect to the control of the duration of ictal events. Another prediction derived from the bistable dynamics, the possibility of aborting absence seizures by means of counter-controlled electrical stimulation, is also discussed in the light of current experimental studies. Finally the most recent

P. Suffczynski (✉)
Department of Biomedical Physics, Institute of Experimental Physics, University of Warsaw, Warsaw, Poland

S. Kalitzin
Foundation Epilepsy Institute in The Netherlands (SEIN), Heemstede, The Netherlands

Image Sciences Institute, University Medical Center Utrecht, Utrecht, The Netherlands
e-mail: skalitzin@sein.nl

F. H. Lopes da Silva (deceased)
Swammerdam Institute for Life Sciences, Center of Neuroscience, University of Amsterdam, Amsterdam, The Netherlands

Department of Bioengineering, Instituto Superior Técnico, Lisbon Technical University, Lisbon, Portugal
e-mail: f.h.lopesdasilva@uva.nl

© Springer Nature Switzerland AG 2019
V. Cutsuridis (ed.), *Multiscale Models of Brain Disorders*, Springer Series in Cognitive and Neural Systems 13, https://doi.org/10.1007/978-3-030-18830-6_16

update of the model was carried out to account for the particular properties of the cortical "driver" of SWDs, and the underlying putative role of the persistent Na^+ current of cortical neurons in this process.

Keywords Generalized epilepsies · Absence seizures · Neural mass models · Dynamical neural systems · Bistability · Inter-ictal and ictal distributions · Counter-stimulation · I_h current · Na-persistent current

16.1   Introduction

This chapter gives an overview of the pathophysiology of absence seizures in the light of insights obtained by means of computational modelling associated with recent neurophysiological experimental evidence. Classically, absence seizures are considered the paradigm of primary generalized epilepsies (PGE). They are characterized by a sudden arrest of ongoing behaviour and conscious awareness, while the electroencephalogram (EEG) displays a burst of bilateral oscillations, in human at about 3/s spike-and-wave discharges (SWDs), which have abrupt onset and cessation. The main question that we address in this overview is what is the dynamical process by which the same brain can produce paroxysms of SWDs associated with the arrest of conscious awareness and normal EEG activity associated with normal state of consciousness. This process is considered from the perspective of bistability, meaning that the brain has two stable equilibrium dynamical states.

16.2   The Conceptual Link Between Absence Epilepsy
and Dynamical Complex Nonlinear Systems

This sudden change in brain rhythmic activity typical of absence seizures reminds us of the seminal writings of Mackey and Glass already in 1977 [24], about the conditions under which "oscillations and chaos" can occur in physiological systems that led to the development of the concept of "dynamical diseases" [25]. In short, these authors proposed that "the signature of a dynamical disease is a change in the qualitative dynamics of some observable nature". In essence these changes in dynamics correspond to bifurcations in a complex nonlinear system, which mathematically describes the physiological system. In an earlier paper [19], we proposed that this is what occurs in the thalamocortical system. In this system different types of oscillations occur, depending on the state of a number of parameters that are controlled by neuromodulatory subsystems: normal oscillations as alpha rhythms or sleep spindles and pathological oscillations as SWDs during absence seizures.

Why, how, and when paroxysmal oscillations of epileptic nature occur is difficult to grasp based simply on current knowledge of pathophysiology. This is the consequence of the complexity of factors that jointly regulate the dynamics of a paroxysmal disease. It may appear trivial, but it should be added that an essential feature of this kind of conditions is that for periods of considerable duration, the behavior and brain signs display mostly normal states of activity, in general much longer than the duration of the paroxysmal episodes of abnormal activity. In order to apprehend the underlying dynamics of a paroxysmal disease, as is the case in epilepsy, it is appropriate to resort to dynamical modelling of complex nonlinear systems.

16.3   A Simplified Sketch of the Thalamocortical Neuronal Networks: The Sources of SWDs

Before describing the computational models, it is relevant to sketch the basic structure and physiology of the system being modelled: the thalamocortical system [37] is shown schematically in Fig. 16.1. The system consists essentially of two neuronal populations, cortical and thalamic, which are mutually interconnected by a number of loops. In a simplified way, the thalamic loop is formed by the population of thalamocortical relay (TCR) neurons that project to the population of reticular thalamic (RE) neurons. The latter inhibit TCR neurons by way of $GABA_A$ and $GABA_B$ types of inhibition. TCR neurons receive also external inputs from sensory systems, from the basal forebrain and the brainstem. The cortical network consists essentially of a negative feedback loop formed by interacting populations of pyramidal neurons (PY) and inhibitory interneurons (IN). Pyramidal cells, in addition to projecting to local interneurons, send also excitatory connections to the thalamus both to the TCR and RE populations. In turn, the TCR cells excite both the pyramidal cells and interneurons. Thalamic populations receive also modulating inputs corresponding, among others, to cholinergic activation from the brain stem.

16.4   Modelling Spike-and-Wave Patterns

Several modelling studies addressed the question which mechanisms are responsible for the generation of SWDs. Most of these focused on the thalamus, the brain structure that was then thought to be primarily responsible for the generation of SWDs; thus, Wang [41] developed a single *neuron model* of a thalamic relay neuron that is completely deterministic and may display "strange attractors" typical of a chaotic dynamic state. A *thalamic neuronal network* spatial model was developed by Lytton et al. [23] which shows that the complex nonlinear dynamics of the neuronal network depends critically on the low-voltage activated (LVA or LTS) Ca^{2+}-current I_T that can lead to intrinsic repetitive bursting in TCR neurons. It is also noteworthy that drugs, like ethosuximide, that depress I_T can suppress

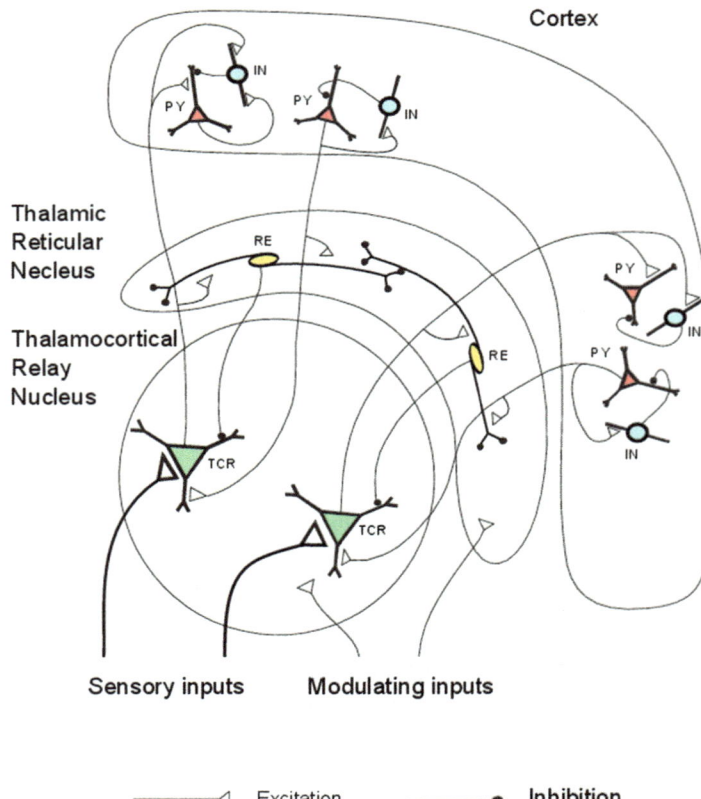

Fig. 16.1 Simplified diagram of the thalamocortical network showing connections between different cell types. Thalamic relay cells (TCR) in the main thalamic nuclei send excitatory input to the reticular nucleus neurons (RE). GABAergic RE cells are mutually connected and send back inhibitory fibres to the TCR cells. Pyramidal cells (PY) in the cortex interact with local interneurons (IN). The cortex sends also recurrent connections to both thalamic relay and the reticular nucleus and receives sensory input via TCR cells. Additionally, thalamic nuclei receive modulatory inputs from the brain stem and basal forebrain structures. Open triangles denote excitatory connections, and filled circles denote inhibitory ones, as indicated in the figure

seizure initiation, although it should be noted that ethosuximide acts also on the persistent Na^+ current (I_{NaP}) and on a Ca^{2+}-activated K^+ current [18]. A *model of cortico-thalamic feedback loops* was developed by Destexhe et al. [7] and Destexhe and Sejnowski [8] in order to investigate the role of cortico-thalamic feedback in promoting SWD oscillations, in contrast with most previous studies where the emphasis was on the role of intra-thalamic mechanisms. The model indicates that cortico-thalamic feedback is of crucial importance and that it acts through excitation of GABAergic RE neurons leading to the recruitment of TCR cell inhibition that can show rebound firing. Among other predictions this model reveals that the upregulation of I_h current, induced by an increase of $[Ca^{2+}]_i$, may mediate the termination of absence seizures.

16.5 Modelling the Dynamics of Absence Seizures

The models described in short above gave insights into some basic neuronal mechanisms underlying the generation of SWDs, but did not address specifically the most relevant issue of the dynamics of this type of epileptic activity, i.e. what causes absence seizures *to start* from a normal background of brain activity *and to stop* after about 5–15 s? This implies that the underlying neuronal system is able to display both kinds of activity – normal EEG and paroxysmal SWDs – i.e. it possesses *bistability*. This means that the dynamics of this system may undergo a transition between those two states, depending on some specific conditions. In other words the system displays a *bifurcation*. With the objective of better grasping this dynamical feature, we developed a computational model using the neural mass approach introduced by Wilson and Cowan [43]. This model consists of four interacting neural populations: TCR, RE, cortical pyramidal cells (PY), and interneurons (INs) integrating synaptic and network properties at the mesoscopic and macroscopic levels (Fig. 16.2) [38, 39]. The model's output signal can display a waxing and waning "spindle-like" oscillation of relatively low amplitude having a spectrum with a peak at approximately 11 Hz, simulating the normal state EEG, or a high amplitude "seizure-like" oscillation at a frequency around 9 Hz that constitutes the limit cycle characteristic of absence seizures in the rat (WAG/Rij or GAERS) (Fig. 16.3).

Thus, for a given set of neuronal parameters, the model is in a "bistable regime" where it may generate both normal and paroxysmal oscillations and spontaneous transitions between these two types of behaviour, depending on exogenous and/or endogenous fluctuations. A representation of the corresponding dynamics in phase space can illustrate what is the main difference between the brain of an epileptic patient with absence seizures and a normal brain, assuming that both share the same basic neuronal networks. Figure 16.4 shows a two-dimensional slice of the excitatory/inhibitory phase space of the dynamical model where the trajectories described by the system are projected. In both cases two attractors are shown: the outer one corresponds to the seizure state and the inner one to the normal state. The essential difference between both is that in the normal case, the two attractors are kept well apart on both sides of the "separatrix"; this is not the case in the epileptic brain. In the latter some parameters differ from the normal situation (e.g. the characteristics of some ion currents in a specific group of neurons, as discussed below). This is expressed by a deformation of the attractors such that the distance between the outer and inner attractors is much smaller. This implies that even random fluctuations, for example, an increase in the power of an external drive, can cause the trajectory to cross the "separatrix", leading to a seizure. This illustrates

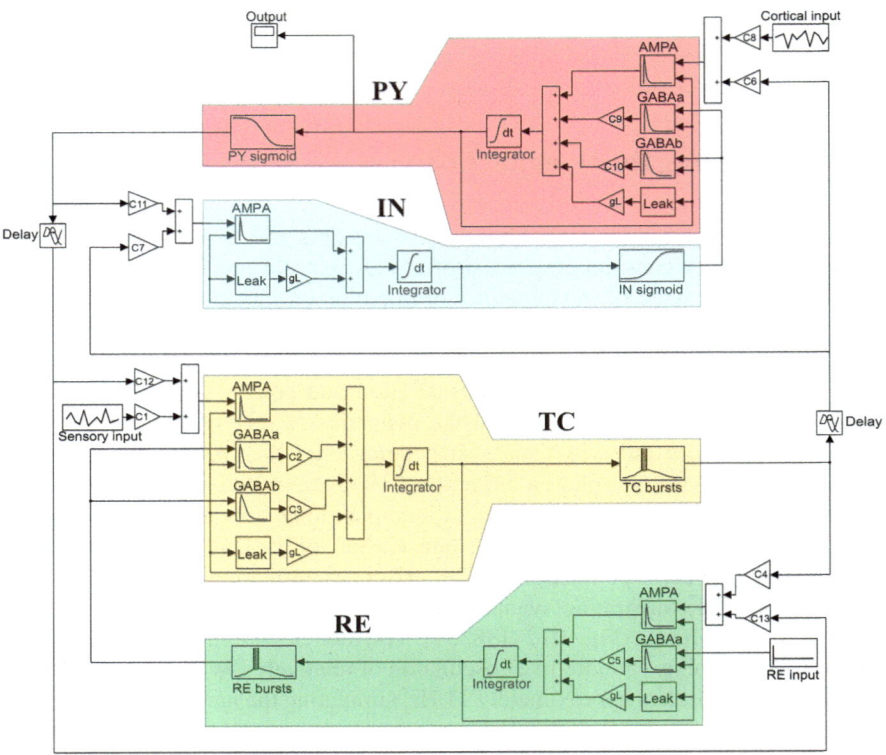

Fig. 16.2 Diagram of the thalamocortical network model. Four neuronal populations are connected according to anatomical connections. Each population is described by synaptic current responses, which are integrated giving rise to mean membrane potential of a population. Population voltage is transformed into population firing by nonlinear spike- (cortex) or burst (thalamus)-generating mechanism. The population of thalamocortical cells (TC, yellow) receives sensory input, while cortical pyramidal cells (PY, red) receive cortical input from other cortical areas. Mutual inhibition between reticular cells (RE, green) is represented by RE inhibitory input. Mean membrane potential of the PY population represents the model output. (Adapted with permission from Suffczynski et al. [38])

the concept that these absence seizures can be caused by random fluctuations of some variables and that their occurrence follows a stochastic process. This feature is an important prediction of the dynamical model. The model was originally aimed to reproduce seizure-like activity in the rat, but with minor modification of some parameters (namely, a reduction of GABA-A conductance and an increase of GABA-B conductance), the model may also account for human SWDs. Comparison of the model output with EEG signals recorded in an epileptic patient is shown in Fig. 16.5.

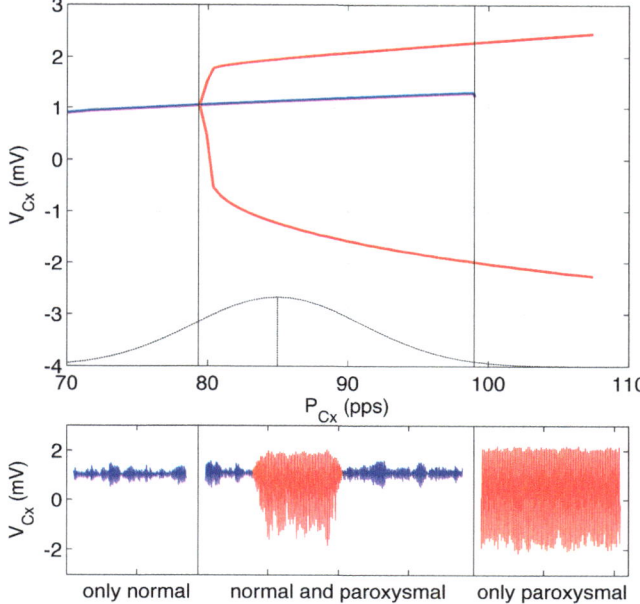

Fig. 16.3 Upper panel: Bifurcation diagram for the thalamocortical population model with cortical P_{Cx} input as control parameter on the x-axis and cortical membrane potential on the y-axis. Solid lines denote stable steady state solutions: fixed point (blue) and limit cycle (red). Thin vertical black lines denote the control parameter space borders at which bifurcations occur and which divide the plot into three different regions. For low P_{Cx} values (on the left side) only fixed point exists. For extreme P_{Cx} values (on the right side), only a limit cycle is present. For intermediate P_{Cx} values, between black vertical solid lines, a bistable region is present in which fixed point solutions coexist with periodic oscillations. Dotted lines below the plot represent the probability distribution of the P_{Cx} input and show that when the mean of the P_{Cx} input is in the bistable region, fluctuations may reach neighbouring regions, what leads to change of a steady state. Lower panel: examples of the model output for three different mean values of the P_{Cx} input located in different parameter space regions. For the mean P_{Cx} located in the mono-stable fixed point region, only normal behaviour is observed. For the mean P_{Cx} in the bistable domain, transitions between normal and paroxysmal activity occur. For the mean P_{Cx} in mono-stable limit cycle region, only paroxysmal activity is present. Colour of the signals corresponds to different types of steady states shown in the upper plot. Time between the ticks is 5 s

16.6   Predictions of the Dynamical Model of Absence Seizures

The model described above leads to two main predictions.

The *first prediction* concerns the probability distributions of inter-ictal intervals and of ictal durations. If the ictal events occur randomly, the intervals between these events are exponentially distributed (Fig. 16.6). This is the special case of the gamma distribution that is described by the following expression, $y = C x^{\alpha-1} e^{-x/\beta}$, where C is a normalization constant and the distribution's parameters are α,

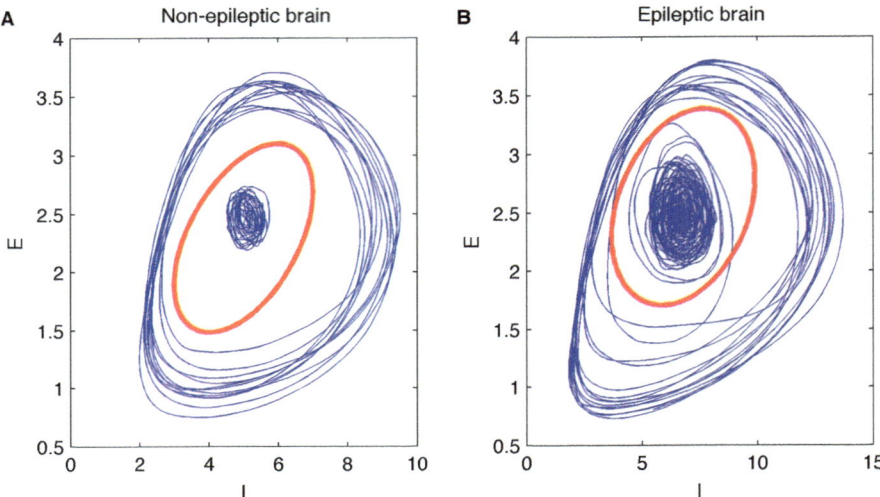

Fig. 16.4 Phase portraits of the model trajectories showing differences between non-epileptic (**a**) and epileptic (**b**) brains. Excitatory cortical activity (**E**) is plotted along the y-axis and inhibitory activity (I) along the x-axis. On each plot fixed point and limit cycle solutions (blue lines) and a boundary between them ("separatrix", red line) are shown. In non-epileptic brain the distance between the fixed point behaviour and the boundary is large, and the trajectory doesn't cross spontaneously into the limit cycle domain. In epileptic brain the distance between trajectories and the boundary is relatively small and the trajectory may cross between domains due to fluctuations in E or I, leading to epileptic transitions. (Adapted from Lytton [22]. And from: Lopes da Silva et al. [19])

the shape parameter, and β, *the scale parameter*. In the case of the exponential distribution, $\alpha = 1$. For the cases where $\alpha > 1$, the process is not completely random, as when some extra number of degrees of freedom has to be taken into account, what is described more explicitly below. In the cases where $\alpha < 1$, the process may correspond to a random walk; this latter model has been proposed in neurophysiology by Gerstein and Mandelbrot [11] to describe how a neuron reaches the firing threshold when the weighted sum of incoming excitatory and inhibitory inputs upon the membrane potential attains a threshold, a process that is akin to a random walk. This may be assumed to occur also with respect to the way that the equivalent electrical potential of a neuronal mass reaches the "separatrix" between two dynamic attractors. Also in this case, successive inter-ictal intervals are not fully independent as the ictal events are clustered, but they still have random duration.

Indeed according to the *first part of the first prediction*, concerning the distribution of *inter-ictal* intervals, the latter follows a gamma distribution with $\alpha \leq 1$. This model prediction was verified experimentally (Fig. 16.6) in WAG/Rij rat and also in GAERS, in human absences, and in an in vitro model [39].

The second part of the first prediction that the same gamma distribution with $\alpha \leq 1$ would apply also to ictal durations, however, was found in a group of

Model simulation

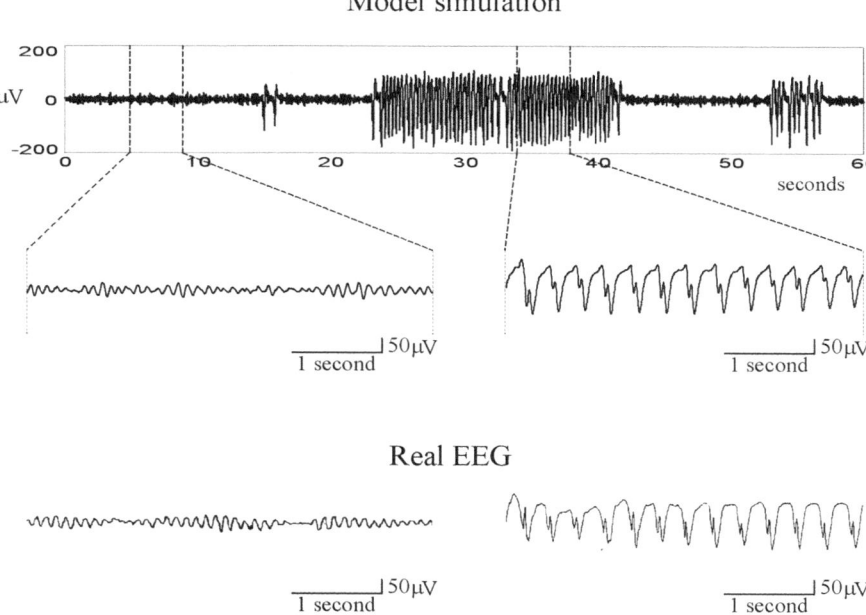

Real EEG

Fig. 16.5 Comparison between the model output and real EEG signals in a human subject. In the upper panel, a simulated epoch of the 60s with two seizure episodes is shown. In the middle panel, enlargements of normal and seizure-like activity are shown. In the bottom panel examples of human EEG during alpha oscillations (on the left) and an absence seizure (on the right) are shown. In order to simulate EEG signals, the sum of postsynaptic currents in both cortical populations was taken as a model output. (Adapted from: Lopes da Silva et al. [20])

experimental rats as shown in Fig. 16.6 that had been treated with vigabatrin, a GABA transaminase inhibitor [4], but this was not supported by experimental observations in other groups of untreated rats. This was, at first, a surprising result; in any case it indicated that the model is missing something essential. The finding that the α-parameter was consistently larger than unity in the gamma distribution of the duration of ictal periods in untreated rats suggests the possible involvement of extra dynamic degrees of freedom in the process of termination of seizures. Therefore, we hypothesize that the deviation of α from unity could be accounted for by the action of "activity-dependent" mechanisms, namely, by "homeostatic mechanisms of regulation of neuronal excitability". The latter might be affected by the treatment of the rats with vigabatrin. In this way we explored a number of possible mechanisms by extending the neural mass model of Suffczynski et al. [38] described above. Thus, Koppert et al. [15] were able to demonstrate in this extended model that the introduction of the hyperpolarization activated inward current I_h, in the pyramidal population, was able to affect the duration of the limit cycle, i.e. the termination of SWDs. In this way the α-parameter associated with the seizure duration distribution increases significantly with increasing I_h channel

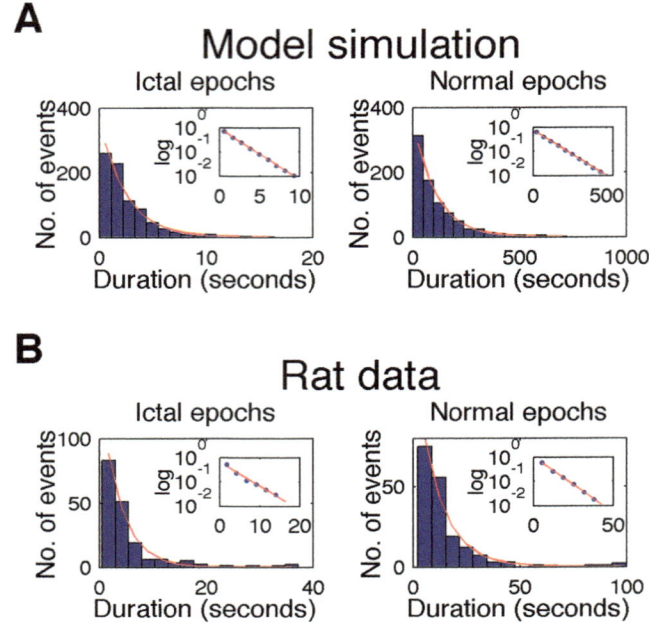

Fig. 16.6 Distributions of the durations of ictal and inter-ictal epochs obtained in the model (**a**) and in the WAG/Rij rat in vivo (**b**). Histograms with experimental data that represent the relative number of events in each interval are shown. Exponential distributions are indicated by red lines and also on the log plots in the inset. Note the similarity between the model and the real data. (Adapted from: Suffczynski et al. [38])

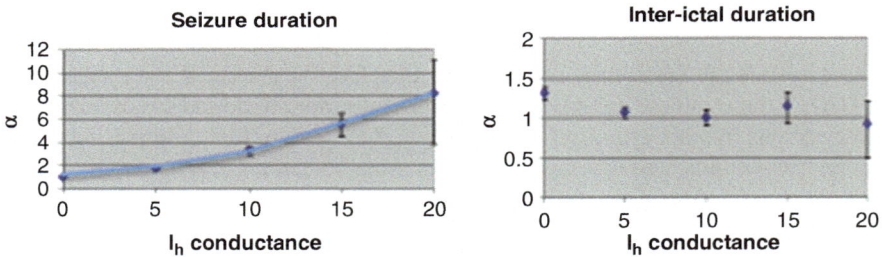

Fig. 16.7 α-parameter values of the gamma distribution of durations of ictal and interictal periods. Increase of I_h current conductance leads to an increase of α above unity for ictal epochs distribution (left) but doesn't affect distributions of inter-ictal epochs (right). (Adapted from Koppert et al. [15])

conductance, while a similar increase of the I_h conductance does not affect the distribution of the duration of inter-ictal periods (Fig. 16.7). This change in the α-parameter characteristic of the distribution of the duration of ictal periods is in agreement with the new experimental findings in untreated rats, reported above.

A *second prediction* is that SWDs may be annihilated by well-timed pulses, since the system is bistable [16, 38]. Indeed some studies appear to support the proposition that counter-stimulation can abort paroxysms of SWDs in experimental animals. Osorio and Frei [31] showed that single DC pulses were effective in aborting generalized seizures in rats. Also it has been shown that brief sensory stimulation delivered at the onset of absence seizure in humans may be effective in arresting a seizure [35].

In a number of studies, a variety of experimental electrical stimulation protocols have been used to control SWDs. Although these experimental manipulations operate in different ways compared to those presented above, i.e. by modulating coarse network parameters, rather than by means of activating a switch between coexisting dynamical states, it is interesting to consider these studies briefly here. Berényi et al. [2] used a closed-loop low-frequency transcranial electrical stimulation (TES) in a rodent model of absence seizures; a short series of electric pulses triggered by the SWDs (at a frequency about 6 Hz) was able to shorten the duration of the SWD burst in an intensity-dependent way. Kozák and Berenyi [17] extended the latter study, using in essence the same protocol with the aim of determining whether "on-demand TES treatment of epileptic seizures" over an extended period of time up to 4 months would have a long-term therapeutic effect, but it did not. This is not surprising taking into consideration that absence seizures are essential manifestation of dynamical bistability. Aborting seizure episodes by temporarily manipulating network parameters does not alter the general dynamics for the occurrence of seizures. A similar paradigm was used by Van Heukelum et al. [40], following Lüttjohaan and van Luijtelaar [21], but now applying high-frequency stimulation (HFS = 130 Hz) to the somatosensory cortex, also in a closed-loop fashion in the same experimental rats; similarly, they reported that this form of stimulation could significantly shorten the duration of SWDs and reduce the number of SWDs, but the SWD duration returned to baseline values in the post-closed-loop stimulation period. This fact is suggestive that the main effect of the HFS is the modulation of cortical excitability in general. Furthermore this study did not reveal whether HFS effects were site specific.

16.7 Where Is the "Driver Network" of SWDs? Experimental Approaches

Variation of model's parameters in our neural mass model [38] showed that, depending on the parameter setting, the SWDs could be initiated in the thalamus or in the cortex. This result does not resolve the enduring discussion as to where SWDs may originate primarily in the cortex or in the thalamus or as Avoli [1] in a striking title expressed "a brief history on the oscillating roles of thalamus and cortex in Absence seizures". This discussion is still very much alive. As mentioned by Sorokin et al. [36], it is a challenging issue to isolate the thalamic and cortical contributions with respect to the primary origin of SWDs by means of lesions, pharmacology, and electrical stimulations. With the aim of throwing light in this discussion, we used an alternative methodology: we applied nonlinear association analysis to local EEGs (local field potentials, LFPs) recorded simultaneously from a large number of cortical and thalamic sites in awake freely moving WAG/Rij rats, enabling us also to estimate short time delays between associated signals [28, 29]. This approach led to the finding that there exists a cortical site where SWDs are initiated in the perioral area of the somatosensory cortex that was called the "cortical driver", from where these propagate with short time delays to other cortical and thalamic sites (Fig. 16.8). This pattern is very consistent within the initial <500 ms of a SWD burst, but the SWD activity becomes bidirectional thereafter. Interestingly, the "driving" function of this cortical site was amply confirmed by new experiments where it was shown that the anti-absence drug ethosuximide was able to suppress SWDs only if applied locally in this peri-oral area in freely moving GAERs, but not if applied in other cortical areas [26]. Most interesting Polack et al. [34], using intracellular recordings, showed in GAERS that SWDs are initiated in layer 5/6 of the same peri-oral area of the somatosensory cortex. Furthermore the ionic mechanism underlying SWDs was shown to involve an interaction between glutamatergic pyramidal neurons, likely amplified by a persistent voltage-gated Na^+ current (I_{NaP}), and inhibitory interneurons that limit the firing of the pyramidal neurons; the resulting hyperpolarization activates the hyperpolarization-activated cationic I_h current, such that rhythmic SWDs ensue [6]. At the molecular level an additional interesting finding was that in the same area of the "neuronal driver of SWDs" an upregulation of mRNA coding for two sub-units of voltage-gated Na^+ channels, that are likely responsible for an enhancement of the I_{NaP} was found [3, 9, 14]. In short, these experimental animal models allowed, not only to find and characterize, at the cellular and molecular levels, the "driver network" that is the primarily responsible for the generation of SWDs, but they also permitted the study of different features of the dynamics of the cortico-thalamic system at multiple scales. In this context an important recent study [36], called attention to the importance of the firing modes of thalamic neurons in the maintenance of SWDs; these authors found that phasic firing is necessary to maintain SWDs oscillations, in contrast to tonic firing. Using optogenetic tools Sorokin et al. [36] demonstrated that switching the firing mode of thalamic cells from phasic to tonic was sufficient to terminate SWDs. These results emphasize the role of the bi-directional cortico-thalamo-cortical system in maintaining SWD oscillations. As these authors note these new findings do not argue against the "cortical driver theory" [29], but suggest further that the thalamo-cortical pathway could act as a "choke point" for SWDs. According to Paz and Huguenard [32] this "choke point" is a site "remote from the initiation site that might be as important as the initial dysfunction".

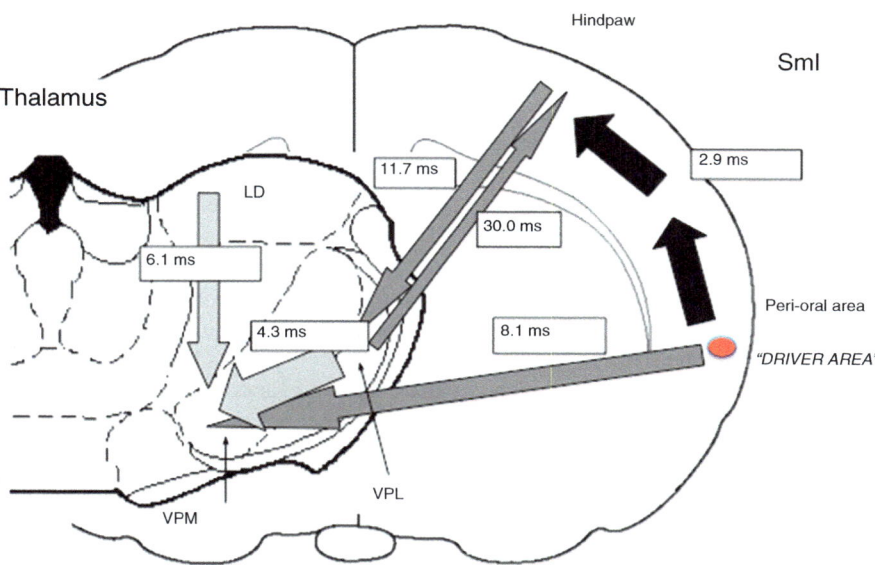

Fig. 16.8 Simplified diagram of the results obtained by Meeren et al. [28] employing nonlinear association (h^2) analysis of the EEG signals recorded simultaneously from multiple cortical and thalamic structures during spontaneous SW discharges in WAG/Rij rats. For the rat shown here, ten absence seizures were analysed. The results shown correspond to the average of the local field potentials recorded during the initial 500 ms of ten bursts of SWDs. The values of averaged time delays in milliseconds are indicated. A consistent cortical *driver* network (indicated by the red circle) was found in the perioral area of the somatosensory cortex (SmI), because this site consistently led the other cortical and thalamic recording sites. Cortico-cortical relationships are represented by black arrows; intra-thalamic by light grey arrows, and cortico-thalamic by dark grey arrows. The thickness of an arrow represents the average strength of the association (i.e. the value of the h^2 index), and the direction of the arrowhead points to the direction of the lagging site. The hindpaw area, for instance, was found to lag by 2.9 ms on average with respect to the *driver* site. Within the thalamus, the laterodorsal (LD) nucleus was found to consistently lead other thalamic sites. The ventroposterior medial (VPM) nucleus was found to lag behind the ventroposterior lateral (VPL) nucleus, with an average time delay of 4.3 ms. Concerning cortico-thalamic interrelationships, the cortical focus site consistently led the thalamus (VPM), with an average time delay of 8.1 ms. Within the somatosensory system of the hindpaw, the (non-focal) cortical site led the thalamic site (VPL) during three of ten seizures; the thalamus led the cortex during one seizure, whereas for the other six seizures, no direction of the delay could be established. (Adapted with permission from Meeren et al. [28])

16.8   Which Properties Make the "Cortical Network" a "Driver' for the Initiation of SWDs? A Modeling Approach

We may reason that some special (patho)physiological properties should be characteristic of neurons in the "driver network" that would reduce the threshold for the bifurcation between normal and SWD activity. In this respect the clue is given by the electrophysiological findings of the Paris group [34, 42] associated with the molecular findings of Klein et al. [14], clearly leading to the assumption that the local pyramidal cells of layers 5/6 in the "cortical driver network" display an abnormal enhancement of the expression of I_{NaP}. To test this hypothesis we added to the original neural mass model of Suffczynski et al. [38, 39] a module to simulate the contribution of this enhanced I_{NaP}. This current tends to amplify small depolarizations [13]. We determined quantitatively what was the effect of the gain of this conductivity on the crossing the threshold or separatrix between the normal and the epileptic attractors. The value of the seizure threshold in the model was quantified as the minimal value of a ramp input to pyramidal cells population necessary for the network to switch to limit cycle oscillations. The result is shown in Fig. 16.9. Part A of this figure shows three sigmoid functions, describing mean firing rate of a neuronal population as a function of the mean membrane voltage, for three different values of the $I_{Na,p}$ conductance: a sigmoid used as reference (ref, in blue), another one (up, in red) shifted to a more negative membrane potential, and a

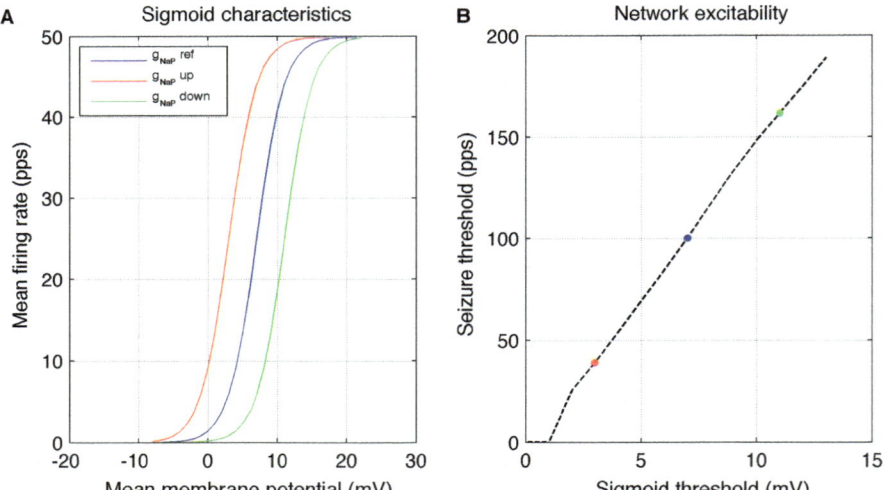

Fig. 16.9 (**a**) Sigmoid functions, describing mean firing rate of a neuronal population (in pulses per second, pps) as a function of mean membrane potential. Three different sigmoids are shown for three different values of the firing threshold, assumed to reflect different conductance of sodium currents, and the extra contribution of the persistent sodium I_{NaP} current. Blue, reference sigmoid; green sigmoid, hypothetical decreased g_{NaP} conductance and increased threshold; red sigmoid, the proposed increased g_{NaP} conductance and decreased threshold. (**b**) Dependence of the network excitability on the pyramidal population sigmoid threshold. The seizure threshold was measured in the model as the additional DC input necessary to trigger a seizure (in pulses per second, pps). The plot shows that as the firing threshold of cortical neurons decreases, e.g. due to an increase of g_{NaP} conductance (red sigmoid in **a**), the seizure threshold decreases; as a result the network becomes more excitable. Each of the three colour points on the broken black line, correspond to each of the sigmoids of the same colour shown in plot **a**

third one (down in green) shifted in opposite direction, for comparison. Figure 16.9b shows the corresponding seizure thresholds; clearly the sigmoid representing the enhanced contribution of the persistent I_{NaP} conductance (up), corresponds to a lower threshold for the transition from normal to seizure-like activity than the original reference value.

Summarizing our main findings, by means of a combined modelling and experimental approach, demonstrate that transitions from the normal to the seizure (SWD) state are controlled by a random walk process that operates between coexisting normal and seizure dynamical states separated by a boundary. Such transitions are noise induced and don't require parameter changes. After crossing the boundary of limit-cycle oscillation, an activity dependent process, moves the system back to the normal state. Accordingly, seizure termination is dependent on a change in network parameters. It is interesting to briefly discuss these results in the light of other modelling studies, in particular in the framework of the concept of multi-stability.

In some other models the transitions from normal state to SWD are mediated by parameter changes [5, 33] or can be triggered either by noise or by a parameter change, depending on whether the model is in the bistable region or not [27]. While seizures caused by parameter change may be induced by pharmacological manipulation, this is not the case for spontaneously occurring episodes. Goodfellow et al. [12] using a spatiotemporal neural mass model suggested bistable switching between low amplitude desynchronized state and synchronous SWDs, governed by intermittent dynamics rather than by noise. The distribution of the durations of seizure-free, and seizure epochs predicted by this model, however, has not yet been confirmed experimentally. Milton et al. [30] approached the dynamical properties of neural microcircuits responsible for absence epilepsy focusing on the concept that time delays within these circuits can play an important role in the dynamics, since these microcircuits can generate multi-stability depending on their internal delays. In such multi-stable system absence seizures correspond to transients associated with transitions between stable attractors. In this context the notion of "multi-stability" means that two or more attractors can coexist, and transient dynamics can occur due to delay-induced transient oscillations. Hence the concept of multi-stability in absence epilepsy appears to be relevant. The various proposed mechanisms that may produce paroxysmal changes in dynamics, however, require further experimental validation.

16.9 Conclusions

In this paper we made an attempt to put together modelling and experimental studies carried out with the objective of obtaining insight into the brain mechanisms underlying absence epilepsy. We elaborated further on our own previous investigations regarding both research fields in order to highlight the importance of the integration of new experimental and modelling studies. Thus the original model of cortico-thalamo-cortical networks [38] was extended in two ways: first, by Koppert et al. [15] adding an homeostatic mechanism of regulation of neuronal excitability (the I_h current) that can account for the termination of a burst of SWDs, and thus for the transition from the limit cycle back to the normal state; second, in the present work, by introducing a module representing the persistent Na^+ current (I_{NaP}) that was found in the physiological experiments [6] to be strongly enhanced in the pyramidal cells of the cortical area responsible for initiating and driving a burst of SWDs. With these extensions, the basic model can account for the distributions of the duration of both inter-ictal normal periods (gamma distribution with $\alpha \leq 1$), and of ictal periods (gamma distribution with $\alpha > 1$).

A central feature elicited by our computational model is the coexistence of multiple stable dynamic states for the same set of parameters. This property has been not considered in other models where the basic mechanisms for changing the behaviour of the system are by shifting the parameters [44, 45]. The presence of multiple states, or attractors, however, can be affected by the values of the parameters in the model [38]. This leads to the possibility [19] that epileptic conditions can be due to variety of parameter settings, a circumstance that may be the explanation of why no universal robust biomarker has been found to identify epileptic tissue or condition so far.

Absence epilepsy is a remarkable form of epilepsy since it is a clear manifestation of a dynamic disease. The neural processes underlying the sudden loss of consciousness, that forms the core of the symptomatology of absence seizures, have intrigued a long series of clinical and experimental neuroscientists, who have offered several explanatory theories, which were the object of lively controversies (for an interesting overview of the latter, see Avoli [1]). According to our point of view, an understanding of those neural processes has necessarily to account for the fine dynamics of absence seizures, with the characteristic short ictal paroxysms and the relative long periods where brain activity and behaviour are normal (inter-ictal periods). This is why we focus on a model that can catch these dynamics and not only the generation of SWDs, as such, with their characteristic waveforms. With this aim we followed the quote of the famous physicist Richard Feynman (in Wikiquote [10]) "What I cannot create, I do not understand". Therefore we *created* a computer model in order to better *understand* the basic neuronal phenomena responsible for the dynamics of absence seizures.

References

1. Avoli M (2012) A brief history on the oscillating roles of thalamus and cortex in absence seizures. Epilepsia 53(5):779–789
2. Berényi A, Belluscio M, Mao D, Buzsáki G (2012) Closed-loop control of epilepsy by transcranial electrical stimulation. Science 337:735–737
3. Blumenfeld H, Klein JK, Schridde U, Vestal M, Rice T, Khera DS, Bashyal C, Giblin K, Paul-Laughinghouse C, Wang F, Phadke A, Mission J, Agarwal RK, Englot DJ, Motelow J, Nerseyan H, Waxman SG, Levin AR (2008) Early treatment suppresses the development of spike-wave epilepsy in a rat model. Epilepsia 49(3):400–409

4. Bouwman BM, Suffczynski P, Midzyanovskaya IS, Matis E, van den Broek PLC, van Rijn CM (2007) The effects of vigabatrin on spike and wave discharges in WAG/Rij rats. Epilepsy Res 76:34–40
5. Breakspear M, Roberts JA, Terry JR, Rodrigues S, Mahant N, Robinson PA (2006) A unifying explanation of primary generalized seizures through nonlinear brain modeling and bifurcation analysis. Cereb Cortex 16:1296–1313
6. Chipaux M, Charpier S, Polack PO (2011) Chloride-mediated inhibition of the ictogenic neurons initiating genetically-determined absence seizures. Neuroscience 192:642–651
7. Destexhe A, Contreras D, Steriade M (1998) Mechanisms underlying the synchronizing action of corticothalamic feedback through inhibition of thalamic relay cells. J Neurophysiol 79:999–1016. [PubMed: 9463458]
8. Destexhe A, Sejnowski TJ (2003) Interactions between membrane conductances underlying thalamocortical slow-wave oscillations. Physiol Rev 83:1401–1453
9. Dezsi G, Ozturk E, Stanic D, Powell KL, Blumenfeld H, O'Brien TJ, Jones NC (2013) Ethosuximide reduces epileptogenesis and behavioral comorbidity in the GAERS model of genetic generalized epilepsy. Epilepsia 54(4):635–643
10. Feynman, R Wikiquote
11. Gerstein GL, Mandelbrot B (1964) Random walk models for the spike activity of a single neurons. Biophys J 4:41–68
12. Goodfellow M, Schindler K, Baier G (2011) Intermittent spike-wave dynamics in a heterogeneous, spatially extended neural mass model. NeuroImage 55(3):920–932
13. Johnston D, Wu SMS (1995) Foundations of cellular neurophysiology. The MIT Press, Cambridge, MA
14. Klein JP, Khera DS, Nersesyan H, Kimchi EY, Waxman SG, Blumenfeld H (2004) Dysregulation of sodium channel expression in cortical neurons in a rodent model of absence epilepsy. Brain Res 1000:102–109
15. Koppert MMJ, Kalitzin SN, Lopes da Silva FH, Viergever MA (2011) Plasticity-modulated seizure dynamics for seizure termination in realistic neuronal models. J Neural Eng 8:1–11
16. Koppert MMJ, Kalitzin SN, Velis D, Lopes da Silva FH, Viergever MA (2013) Reactive control of epileptiform discharges in realistic computational neuronal models with bistability. Int J Neural Syst 23(1). (No 1230032)):1–10
17. Kozák G, Berenyi A (2017) Sustained efficacy of closed loop electrical stimulation for long-term treatment of absence epilepsy in rats. Sci Rep 7(6300):1–9
18. Leresche N, Parri HR, Erdemli G, Guyon A, Turner JP, Williams SR, Asprodini E, Crunelli V (1998) On the action of the anti-absence drug ethosuximide in the rat and cat thalamus. J Neurosci 18:4842–4853
19. Lopes da Silva FH, Blanes W, Kalitzin SN, Parra J, Suffczynski P, Velis DN (2003) Dynamical diseases of brain systems: different routes to epileptic seizures. IEEE Trans Biomed Eng 50(5):540–548
20. Lopes da Silva FH et al (2003) Epilepsies as dynamical diseases of brain systems: basic models of the transition between normal and epileptic activity. Epilepsia 44(Suppl. 12):72–83
21. Lüttjohann A, van Luijtelaar G (2013) Thalamic stimulation in absence epilepsy. Epilepsy Res 106:136–145
22. Lytton WW (2008) Computer modeling of epilepsy. Nat Rev Neurosci 9:626–637
23. Lytton WW, Contreras D, Destexhe A, Steriade M (1997) Dynamic interactions determine partial thalamic quiescence in a computer network model of spike-and-wave seizures. J Neurophysiol 77:1676–1696
24. Mackey MC, Glass L (1977) Oscillation and chaos in physiological control systems. Sci New Ser 197(4300):287–289
25. Mackey MC, Milton JM (1987) Dynamical diseases. Ann N Y Acad Sci 504:16–32
26. Manning JPA, Richards DA, Leresche N, Crunelli V, Bowery NG (2004) Cortical-area-specific block of genetically determined absence seizures by ethosuximide. Neuroscience 123:5–9

27. Marten F, Rodrigues S, Suffczynski P, Richardson MP, Terry JR (2009) Derivation and analysis of an ordinary differential equation mean-field model for studying clinically recorded epilepsy dynamics. Phys Rev E 79(2 Pt 1):021911
28. Meeren HK, Pijn JP, Van Luijtelaar EL, Coenen AM, Lopes da Silva FH (2002) Cortical focus drives widespread corticothalamic networks during spontaneous absence seizures in rats. J Neurosci 22:1480–1495
29. Meeren H, van Luijtelaar G, Lopes da Silva F, Coenen A (2005) Evolving concepts on the pathophysiology of absence seizures: the cortical focus theory. Arch Neurol 62:371–376
30. Milton J, Jianhong W, Campbell SA, Bélair J (2017) Outgrowing neurological diseases: microcircuits, conduction delay and childhood absence epilepsy. In: Érdi P, Sen Bhattacharya B, Cochran AL (eds) Computational neurology and psychiatry. Springer International Publishing, Cham, pp 11–47
31. Osorio I, Frei MG (2009) Seizure abatement with single dc pulses: is phase resetting at play? Int J Neural Syst 19(3):149–156
32. Paz JT, Huguenard JR (2015) Microcircuits and their interactions in epilepsy: is the focus out of focus? Nat Neurosci 18:351–359.
33. Robinson PA, Rennie CJ, Rowe DL (2002) Dynamics of large-scale brain activity in normal arousal states and epileptic seizures. Phys Rev E 65:041924
34. Polack PO, Guillemain I, Hu E, Deransart C, Depaulis A, Charpier S (2007) Deep layer somatosensory cortical neurons initiate spike-and-wave discharges in a genetic model of absence seizures. J Neurosci 27:6590–6599
35. Rajna P, Lona C (1989) Sensory stimulation for inhibition of epileptic seizures. Epilepsia 30:168–174
36. Sorokin JM, Davidson TJ, Frechette E, Abramian AM, Deisseroth K, Huguenard JR, Paz JT (2017) Bidirectionsl control of generalized epilepsy networks via rapid real-time switching of firing mode. Neuron 93:149–210
37. Steriade M, McCormick DA, Sejnowski TJ (1993) Thalamocortical oscillations in the sleeping and aroused brain. Science 262:679–685
38. Suffczynski P, Kalitzin SN, Lopes da Silva FH (2004) Dynamics of nonconvulsive epileptic phenomena modeled by a bistable neuronal network. Neuroscience 126:467–484
39. Suffczynski P, Lopes da Silva FH, Parra J, Velis DN, Bouwman BM, van Rijn CM, van Hese P, Boon P, Khosravani H, Derchansky M, Carlen P, Kalitzin SN (2006) Dynamics of epileptic phenomena determined from statistics of ictal transitions. IEEE Trans Bio Med Eng 53(3):524–532
40. Van Heukelum S, Kelderhuis J, Janssen P, van Luijtelaar G, Lüttjohann A (2016) Timing of high-frequency stimulation in a genetic absence model. Neuroscience 324:191–201
41. Wang XJ (1994) Multiple dynamical modes of thalamic relay neurons: rhythmic bursting and intermittent phase-locking. Neuroscience 59:21–31
42. Williams MS, Altwegg-Boussac T, Chavez M, Lecas S, Mahon S, Charpier S (2016) Integrative properties and transfer function of cortical neurons initiating s=absence seizures in a rat genetic model. J Physiol 594(22):6733–6751
43. Wilson HR, Cowan JD (1972) Excitatory and inhibitory interactions in localized populations of model neurons. Biophys J 12:121–124
44. Wendling F, Bartolomei F, Bellanger JJ, Chauvel P (2002) Epileptic fast activity can be explained by a model of impaired GABAergic dendritic inhibition. Eur J Neurosci 15:1499–1508
45. Wendling F, Hernandez A, Bellanger JJ, Chauvel P, Bartolomei F (2005) Interictal to ictal transition in human temporal lobe epilepsy; insights from a computational model of intracerebral EEG. J Clin Neurophysiol 22:343–356

Chapter 17
Data-Driven Modeling of Normal and Pathological Oscillations in the Hippocampus

Ivan Raikov and Ivan Soltesz

Abstract Epilepsy is a disorder caused by abnormalities at all levels of neural organization that often have complex and poorly understood interactions. Physiologically detailed computational models provide valuable tools for evaluation of possible clinical treatments of neural disorders because every parameter can be changed and many experimentally inaccessible variables can be observed. We present our recently developed full-scale model of the CA1 subfield in the rodent hippocampus and highlight its role in the study of biophysical neural oscillations, which are important biomarkers of cognitive processes as well as abnormal neural dynamics in epilepsy. This model provides an integrative framework that unifies experimentally derived knowledge about the hippocampus on multiple scales and can yield insight into the neurophysiological mechanisms underlying the dynamical regimes of the brain. Such a framework can be useful in studying cellular mechanisms of multitarget pharmacological treatments of neural disorders.

Keywords Hippocampus · CA1 · Theta oscillations · Gamma oscillations · Epilepsy · High-frequency oscillations · Fast ripples

17.1 Introduction

Epilepsy is a spectrum disorder that is characterized by a wide range of unpredictable seizures and is often associated with other cognitive, psychiatric, and sleep disorders [26]. Epileptic seizures are caused by abnormal neuronal excitability, which is influenced by many factors. Gene mutations in ion channels can cause disruptions in neuronal dynamics that can result in cognitive deficits [45]. At the synapse level, changes in the subunit composition of $GABA_A$ receptors [51] and in chloride transporter dynamics [23] can facilitate seizure occurrence. Studies

I. Raikov (✉) · I. Soltesz
Department of Neurosurgery, Stanford University, Stanford, CA, USA
e-mail: iraikov@stanford.edu; isoltesz@stanford.edu

© Springer Nature Switzerland AG 2019
V. Cutsuridis (ed.), *Multiscale Models of Brain Disorders*, Springer Series in
Cognitive and Neural Systems 13, https://doi.org/10.1007/978-3-030-18830-6_17

of temporal lobe epilepsy have revealed that at the circuit level, specific classes of neurons are involved in seizure generalization and comorbidities [11], while at the network level, abnormal spiking activity emerges concurrently in mesial and neocortical temporal regions [34] and optogenetic manipulation of inhibitory cerebellar neurons results in a decrease in the duration of temporal lobe seizures [35]. Taken together, these results suggest the existence of mechanisms that both destabilize local circuits and allow the spread of epileptogenic activity to multiple brain regions well before a seizure occurs.

Conceptually, seizures are thought to result from an inability of intrinsic feedback mechanisms to regulate the coupling strength between interconnected inhibitory and excitatory neurons. Under normal conditions, neural oscillatory activity remains well-regulated when exposed to diverse sensory stimuli, which implies a robust feedback path that maintains a tight balance between excitation and inhibition [3]. Under epileptic conditions, poorly regulated feedback paths lead to a high degree of synchronization that is manifested by seizures [38].

These intrinsic feedback mechanisms arise from specific cellular and synaptic physiological properties of neurons, as well as anatomical and functional connectivity. However, establishing causality between the physiology of neurons and functional pathways of brain circuits is difficult in most experimental work, as data are typically obtained for only a relatively small number of cells or field potentials at a few locations, often over a short time course. The multiscale nature of epileptogenic processes can be addressed by biophysically detailed computational models [40], which enable the selective manipulation of experimentally inaccessible aspects of neural function by removing or adding certain types of neurons, changing ion channel properties, or modifying synaptic connections.

In the next sections, we review the role of network oscillations as a biomarker for epilepsy and the construction of multiscale computational models that exhibit physiological oscillatory behavior.

17.2 Pathological Oscillatory Dynamics as a Biomarker for Epilepsy

The hippocampus has been experimentally observed to generate three major types of oscillatory field potentials: theta rhythms, gamma rhythms, and sharp wave-ripple (SWR) complexes. These oscillatory patterns are of behavioral and cognitive importance [13, 14, 16] and are frequently disrupted or otherwise altered during seizures and other pathological conditions. Among the oscillations of particular interest to epilepsy research are high-frequency oscillations (HFOs). Normal HFOs include ripple oscillations (approximately 100–200 Hz local field oscillations), which are part of the SWR complex: they are initiated by large-amplitude sharp waves that reflect the postsynaptic effects of synchronous discharges of groups of pyramidal cells in the CA3 or CA2 subfield of the hippocampus [14] and are manifested in the CA1 subfield but are also found in the subiculum and entorhinal cortex [21]. SWRs induce cortical synaptic modifications that are involved in memory consolidation [15] and it is thought that the high degree of synchrony of the firing of the participating pyramidal cells is an efficient mechanism to transfer episodic memory traces to the neocortex that comes at the cost of susceptibility to epileptiform discharges in pathological conditions [14].

In rats that exhibit recurrent spontaneous seizures, HFOs with frequencies greater than 200 Hz occur in the dentate gyrus, in the CA1 and CA3 subfields of hippocampus, as well as in the subiculum and entorhinal cortex [8]. These HFOs, called fast ripples, are considered to be pathological events based on results that associated them with seizure onset and correlated a greater number of fast ripple-generating sites with a higher rate of seizures [6, 7, 10, 24, 29, 32, 53]. These results suggest that the cellular and network mechanisms that generate pathological HFOs are important targets for clinical intervention [36].

Computational models are a powerful tool for multiscale simulations of neuronal networks and the oscillations they generate and therefore could become a solid test platform for understanding the mechanisms of epilepsy and exploring seizure control and treatment options. In the next section, we review the recent full-scale model of the CA1 network that was developed in the Soltesz lab.

17.3   A Multiscale, Data-Driven Model of Oscillations in the Hippocampus

The hippocampal CA1 area is involved in diverse cognitive tasks including learning, memory, and spatial processing [43]. These cognitive tasks require coordination of neuronal activity reflected by physiological network oscillations, including the theta rhythm [13, 17]. In rodents, hippocampal theta is a 5–10 Hz oscillation in the local field potential (LFP) and neuronal firing patterns [37, 50, 57, 59, 60] that occurs during locomotion and in REM sleep [13]. Recent reports have suggested an intrinsic ability of the CA1 circuitry to generate theta oscillations even in the absence of rhythmic external inputs [2, 28]. However, the multiscale mechanisms that contribute to the generation of the theta rhythm are not well understood [18, 19].

Motivated to explore the mechanisms of hippocampal oscillations in CA1, recent work in the Soltesz lab has resulted in the most detailed to-date quantitative estimate of the cellular and synaptic constituents of the rodent CA1 region and a corresponding full-scale computational model [4, 5]. The CA1 network model incorporated multicompartmental models of all known CA1 neuron types, including pyramidal cells with realistic morphology and eight types of interneurons with simplified morphology, including PV+ basket cells, CCK+ basket cells, bistratified cells, axo-axonic cells, O-LM cells, Schaffer collateral-associated cells, neurogliaform cells, and ivy cells (Fig. 17.1a). Each cell type has its own complement of ion channel

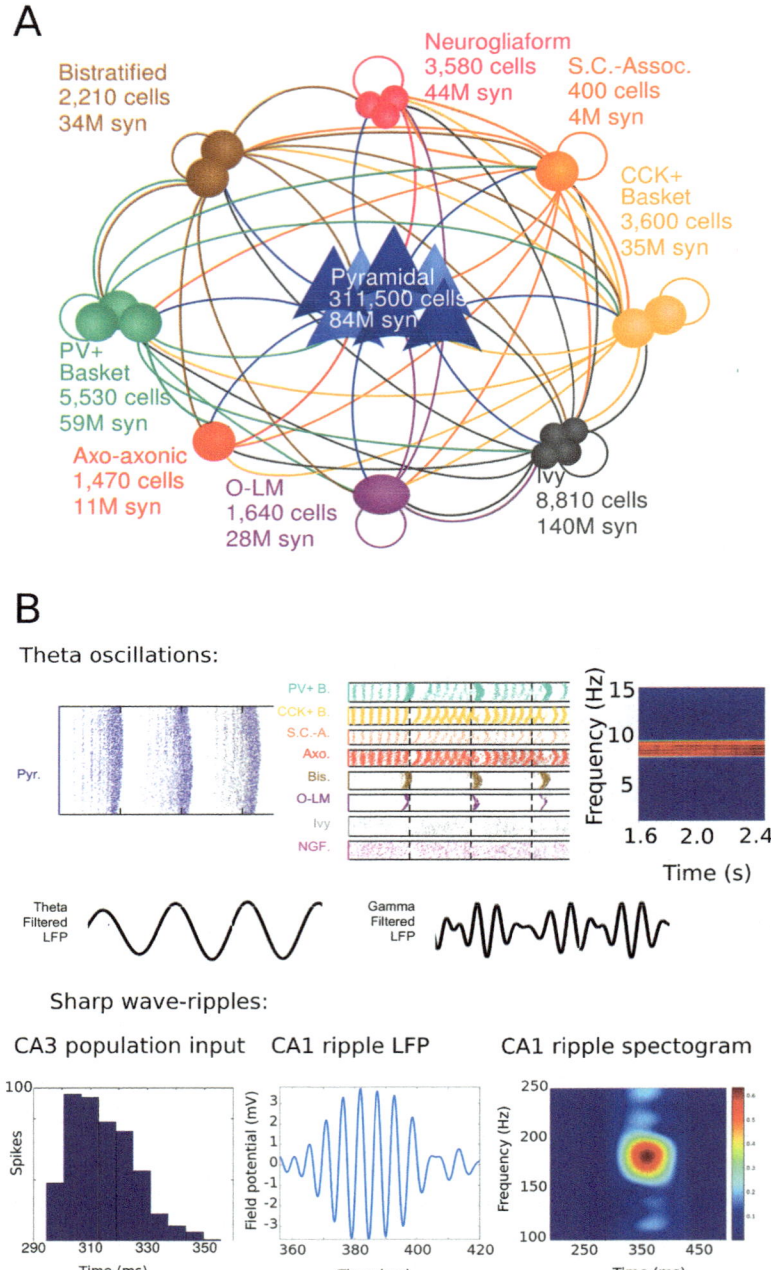

Fig. 17.1 Intrinsically generated network oscillations in a full-scale computational model of the CA1 network. (**a**) Cell types and connectivity structure of the model. (**b**) Top: continuous low frequency asynchronous input from CA3 and ECIII drive theta and gamma frequency oscillations in CA1. Bottom: transient high frequency asynchronous activation of a subset of CA3 inputs results in ripple oscillations in CA1. (Adapted from Bezaire et al. [4])

conductances and thus exhibits realistic intrinsic electrophysiological excitability properties and firing patterns.

The cell populations were organized in 3D with layer depth based on anatomical laminar distributions and randomly distributed horizontal locations. Probability of connection depended on layer and axonal extent of each cell type. Modeled synapse types included AMPA, $GABA_A$, and $GABA_B$. The model contained an implementation of the LFP estimation method described by Schomburg et al. [49], a highly accurate yet computationally efficient approximation.

When excited with tonic inputs at physiologically relevant frequencies, the model network generated spontaneous theta and gamma oscillations, as measured in the LFP signal and spike density function [55], with phase-preferential firing specific to each cell classes present in the model (Fig. 17.1b). Consistent with experimental results [2, 28], these oscillations emerge without explicit patterned or phasic inputs. The model results corresponded well with findings on the differential roles of PV+ basket cells and OLM cells [33]. In addition, the model unexpectedly revealed that interneuronal diversity itself may also be important in theta generation, since replacing all interneurons in the model with fast spiking PV+ basket cells did not result in a theta-generating network, in spite of the key role of PV+ basket cells in hippocampal oscillations.

The model network also displayed gamma oscillations (25–80 Hz), as expected based on in vivo data [20, 50] and in vitro slice data showing 65–75 Hz gamma oscillations arising in response to theta rhythmic network stimulation [12]. The gamma oscillation was phase-locked to the theta rhythm, as it is in vivo [9, 16, 30, 50]. Furthermore, preliminary results indicate that activation of a small group of CA3 afferents consistently yields multi-cycle spike sequences that generate field potentials very similar to ripple oscillations during analogous optogenetic experiments [54] (Fig. 17.1c, d).

The CA1 model results characterize the roles for specific cellular, synaptic, and network components and suggest that the complex hippocampal circuitry maintains the stability of oscillatory mechanisms as the brain operates in diverse behavioral states. It has been theorized that the predisposition for oscillation at theta and gamma frequencies, coupled with phase-preferential firing, may aid information processing by providing order and allowing parallel channels of information [1, 27, 41].

17.4   Outlook

Complex multiscale diseases such as epilepsy require complex multitarget treatments [59]. However, multitarget pharmacological treatment of epilepsy is used without detailed understanding of the interrelated effects of drug combinations [39]. High-level models that describe dynamical processes and seizures in the brain on a macroscopic level do exist [31], but they lack description of ion channel and synaptic dynamics, which makes them unsuitable for quantitatively assessing pharmacological impact. Biophysical models have not been developed to the point of describing several interconnected brain regions but do provide the detail necessary to directly assess pharmacological intervention across temporal and spatial scales.

Our biophysical full-scale computational model of the hippocampal CA1 sub-field provides an integrative framework that unifies experimentally derived knowledge about the hippocampus on multiple scales to recapitulate key behaviorally relevant oscillatory regimes and facilitates the manipulation of cellular, synaptic, and network parameters so as to yield insight into the underlying neurophysiological mechanisms. Combined with models of ion channels and synaptic receptors that are affected by anticonvulsant drugs [44], and in concert with models of other hippocampal and cortical regions [22, 47] and methodologies for connectivity analysis and dendritic modeling [25, 42, 46, 48, 52, 56], this framework will allow researchers an alternative novel tool to model the synergistic effects of different pharmacological treatments on all levels of neural organization.

Acknowledgments The authors are supported by NIH NINDS under award number 1U19NS104590. Access to supercomputers for simulations of the CA1 model was provided by the Extreme Science and Engineering Discovery Environment (XSEDE; NSF grant number ACI-1053575), XSEDE Research Allocation grant TG-IBN140007 to I.S., and the Blue Waters sustained-petascale computing project (supported by NSF Awards OCI-0725070 and ACI-1238993 and the state of Illinois), NSF PRAC Awards 1614622, 1811597 to I.S.

References

1. Akam T, Kullmann D (2012) Efficient communication through coherence requires oscillations structured to minimize interference between signals. PLoS Comput Biol 8:e1002760
2. Amilhon B, Huh C, Manseau F et al (2015) Parvalbumin interneurons of hippocampus tune population activity at theta frequency. Neuron 86:1277–1289
3. Atallah B, Scanziani M (2009) Instantaneous modulation of gamma oscillation frequency by balancing excitation with inhibition. Neuron 62:566–577
4. Bezaire M, Raikov I, Burk K et al (2016) Interneuronal mechanisms of hippocampal theta oscillations in a full-scale model of the rodent CA1 circuit. elife 5:e18566
5. Bezaire M, Soltesz I (2013) Quantitative assessment of CA1 local circuits: knowledge base for interneuron-pyramidal cell connectivity. Hippocampus 23:751–785
6. Bragin A, Benassi S, Kheiri F, Engel J (2011) Further evidence that pathologic high-frequency oscillations are bursts of population spikes derived from recordings of identified cells in dentate gyrus. Epilepsia 52:45–52
7. Bragin A, Engel J, Staba R (2010) High-frequency oscillations in epileptic brain. Curr Opin Neurol 23:151–156
8. Bragin A, Engel J, Wilson C et al (1999) Hippocampal and entorhinal cortex high-frequency oscillations (100–500 Hz) in human epileptic brain and in kainic acid-treated rats with chronic seizures. Epilepsia 40:127–137
9. Bragin A, Jando G, Nadasdy Z et al (1995) Gamma (40–100 Hz) oscillation in the hippocampus of the behaving rat. J Neurosci 15:47–60
10. Bragin A, Wilson C, Engel J (2003) Spatial stability over time of brain areas generating fast ripples in the epileptic rat. Epilepsia 44:1233–1237
11. Bui A, Nguyen T, Limouse C et al (2017) Dentate gyrus mossy cells control spontaneous convulsive seizures and spatial memory. Science 359:787–790

12. Butler J, Mendonca P, Robinson H, Paulsen O (2016) Intrinsic cornu ammonis area 1 theta-nested gamma oscillations induced by optogenetic theta frequency stimulation. J Neurosci 36:4155–4169
13. Buzsaki G (2002) Theta oscillations in the hippocampus. Neuron 33:325–340
14. Buzsaki G (2015) Hippocampal sharp wave-ripple: a cognitive biomarker for episodic memory and planning. Hippocampus 25:1073–1188
15. Buzsaki G (1996) The hippocampo-neocortical dialogue. Cereb Cortex 6:81–92
16. Buzsaki G, Buhl D, Harris K et al (2003) Hippocampal network patterns of activity in the mouse. Neuroscience 116:201–211
17. Buzsaki G, Moser E (2013) Memory, navigation and theta rhythm in the hippocampal-entorhinal system. Nat Neurosci 16:130–138
18. Colgin L (2013) Mechanisms and functions of theta rhythms. Ann Rev Neurosci 36:295–312
19. Colgin L (2016) Rhythms of the hippocampal network. Nat Rev Neurosci 17(4):239–249
20. Colgin L, Moser E (2010) Gamma oscillations in the hippocampus. Physiology 25:319–329
21. Csicsvari J, Hirase H, Czurko A et al (1999) Oscillatory coupling of hippocampal pyramidal cells and interneurons in the behaving rat. J Neurosci 19:274–287
22. Dyhrfjeld-Johnsen J, Santhakumar V, Morgan RJ et al (2007) Topological determinants of epileptogenesis in large-scale structural and functional models of the dentate gyrus derived from experimental data. J Neurophysiol 97:1566–1587
23. Dzhala V, Kuchibhotla K, Glykys J et al (2010) Progressive nKCC1-dependent neuronal chloride accumulation during neonatal seizures. J Neurosci 30:11745–11761
24. Engel JJ, Bragin A, Staba R, Mody I (2009) High-frequency oscillations: what is normal and what is not? Epilepsia 50:598–604
25. Engel JJ, Thompson P, Stern JM et al (2013) Connectomics and epilepsy. Curr Opin Neurol 26:186–194
26. Fisher R, van Emde Boas W, Blume W et al (2004) Epileptic seizures and epilepsy: definitions proposed by the international league against epilepsy (ILAE) and the International Bureau for Epilepsy (IBE). Epilepsia 46:470–472
27. Fries P (2015) Rhythms for cognition: communication through coherence. Neuron 88:220–235
28. Goutagny R, Jackson J, Williams S (2009) Self-generated theta oscillations in the hippocampus. Nat Neurosci 12:1491–1493
29. Jacobs J, Staba R, Asano E et al (2012) High-frequency oscillations (hFOs) in clinical epilepsy. Prog Neurobiol 98:302–315
30. Jensen O, Colgin L (2007) Cross-frequency coupling between neuronal oscillations. Trends Cogn Sci 11:267–269
31. Jirsa V, Stacey W, Quilichini P et al (2014) On the nature of seizure dynamics. Brain 137:2210–2230
32. Jirsch J, Urrestarazu E, LeVan P et al (2006) High-frequency oscillations during human focal seizures. Brain 129:1593–1608
33. Klausberger T, Magill P, Marton L et al (2003) Brain-state- and cell-type-specific firing of hippocampal interneurons in vivo. Nature 421:844–848
34. Kobayashi E, Grova C, Tyvaert L et al (2009) Structures involved at the time of temporal lobe spikes revealed by interindividual group analysis of EEG/fMRI data. Epilepsia 50:2549–2556
35. Krook-Magnuson E, Szabo G, Armstrong C et al (2014) Cerebellar directed optogenetic intervention inhibits spontaneous hippocampal seizures in a mouse model of temporal lobe epilepsy. eNeuro 1:
36. Le Van Quyen M, Khalilov I, Ben-Ari Y (2006) The dark side of high-frequency oscillations in the developing brain. Trends Neurosci 29:419–427
37. Lee M, Chrobak J, Sik A et al (1994) Hippocampal theta activity following selective lesion of the septal cholinergic system. Neuroscience 62:1033–1047
38. Lopes da Silva F, Blanes W, Kalitzin S et al (2003) Dynamical diseases of brain systems: different routes to epileptic seizures. IEEE Trans Biomed Eng 50:540–548
39. Lytton W (2017) Multiscale modeling in the clinic: diseases of the brain and nervous system. Brain Inform 4:219–230

40. Lytton W (2008) Computer modelling of epilepsy. Nat Rev Neurosci 9:626–637
41. Maris E, Fries P, van Ede F (2016) Diverse phase relations among neuronal rhythms and their potential function. Trends Neurosci 39:86–99
42. Morgan R, Soltesz I (2008) Nonrandom connectivity of the epileptic dentate gyrus predicts a major role for neuronal hubs in seizures. Proc Natl Acad Sci USA 105:6179–6184
43. Moser E, Kropff E, Moser M-B (2008) Place cells, grid cells, and the brain's spatial representation system. Annu Rev Neurosci 31:69–89
44. Rogawski M, Cavazos J (2015) Mechanisms of action of antiepileptic drugs. In: Wyllie E (ed) Wyllie's treatment of epilepsy: principles and practice, 6th edn. Wolters Kluwer Health, Philadelphia
45. Rogawski M, Loscher W (2004) The neurobiology of antiepileptic drugs. Nat Rev Neurosci 5:554–564
46. Rubinov M, Sporns O (2010) Complex network measures of brain connectivity: uses and interpretations. NeuroImage 52:2059–2069
47. Santhakumar V, Aradi I, Soltesz I (2005) Role of mossy fiber sprouting and mossy cell loss in hyperexcitability: a network model of the dentate gyrus incorporating cell types and axonal topography. J Neurophysiol 93:437–453
48. Schneider C, Cuntz H, Soltesz I (2014) Linking macroscopic with microscopic neuroanatomy using synthetic neuronal populations. PLoS Comput Biol 10:e1003921
49. Schomburg E, Anastassiou C, Buzsaki G, Koch C (2012) The spiking component of oscillatory extracellular potentials in the rat hippocampus. J Neurosci 32:11798–11811
50. Soltesz I, Deschenes M (1993) Low-and high-frequency membrane potential oscillations during theta activity in CA1 and CA3 pyramidal neurons of the rat hippocampus under ketamine-xylazine anesthesia. J Neurophysiol 70:97–97
51. Sperk G, Drexel M, Pirker S (2009) Neuronal plasticity in animal models and the epileptic human hippocampus. Epilepsia 50(Suppl 12):29–31
52. Sporns O (2012) Discovering the human connectome. MIT Press, Cambridge, MA
53. Staba R, Wilson C, Bragin A et al (2002) Quantitative analysis of high-frequency oscillations (80–500 hz) recorded in human epileptic hippocampus and entorhinal cortex. J Neurophysiol 88:1743–1752
54. Stark E, Roux L, Eichler R et al (2014) Pyramidal cell-interneuron interactions underlie hippocampal ripple oscillations. Neuron 83:467–480
55. Szucs A (1998) Applications of the spike density function in analysis of neuronal firing patterns. J Neurosci Methods 81:159–167
56. Taylor P, Kaiser M, Dauwels J (2014) Structural connectivity based whole brain modelling in epilepsy. J Neurosci Methods 246:51–57
57. Varga C, Golshani P, Soltesz I (2012) Frequency-invariant temporal ordering of interneuronal discharges during hippocampal oscillations in awake mice. Proc Natl Acad Sci 109:E2726–E2734
58. Varga C, Oijala M, Lish J, Szabo GG, Bezaire M, Marchionni I, Golshani P, Soltesz I (2014) Functional fission of parvalbumin interneuron classes during fast network events. Elife 3. https://doi.org/10.7554/eLife.04006
59. Viayna E, Sola I, Di Pietro O, Munoz-Torrero D (2013) Human disease and drug pharmacology, complex as real life. Curr Med Chem 20:1623–1634
60. Ylinen A, Bragin A, Nadasdy Z et al (1995) Sharp wave-associated high-frequency oscillation (200 Hz) in the intact hippocampus: network and intracellular mechanisms. J Neurosci 15:30–46

Chapter 18
Shaping Brain Rhythms: Dynamic and Control-Theoretic Perspectives on Periodic Brain Stimulation for Treatment of Neurological Disorders

John D. Griffiths and Jérémie R. Lefebvre

Abstract Rhythmic, collective activity is a fundamental feature of neural systems. As a result of this, many of the challenges and opportunities involved in developing clinical tools from basic neuroscience knowledge come down to questions about control of dynamic, oscillatory networks. In this chapter we review a range of experimental and theoretical work on control of neural oscillations, in healthy brains and in relation to various clinical conditions. We highlight the main types of qualitative system behaviour that can result from application of periodic stimulation and present a simple case study on this using a mathematical model of rhythmogenesis in thalamocortical circuits. The concepts discussed here may, we hope, help provide some guidelines and principles for the development of future generations of more physiologically and dynamically informed brain stimulation techniques, paradigms, and researchers.

18.1 Background

Synchronous neural activity has been identified as a hallmark of neural communication and information processing. These rhythmic electrical fluctuations, which occur at frequencies spanning several orders of magnitude, have been linked to numerous neurophysiological and cognitive processes, including atten-

---

J. D. Griffiths (✉)
Krembil Centre for Neuroinformatics, Centre for Addiction and Mental Health, Toronto, ON, Canada

Department of Psychiatry, University of Toronto, Toronto, ON, Canada
e-mail: john.griffiths@camh.ca

Jérémie R. Lefebvre
Krembil Research Institute, University Health Network, Toronto, ON, Canada

Department of Mathematics, University of Toronto, Toronto, ON, Canada
e-mail: jeremie.lefebvre@uhnresearch.ca

© Springer Nature Switzerland AG 2019
V. Cutsuridis (ed.), *Multiscale Models of Brain Disorders*, Springer Series in
Cognitive and Neural Systems 13, https://doi.org/10.1007/978-3-030-18830-6_18

tion [1], memory [2], spatial navigation [3, 4], sensory processing [1, 5–7], and a wide variety of perceptual and motor tasks (see [8] and references therein). Abnormal neural synchrony and altered oscillatory activity are hallmarks of many neurological disorders [9] such as Parkinson's [10], dementia [11, 12], stroke [13–15], tumours [16–18], and neurogenic pain [19, 20]. These observations, together with recent technological developments and increasingly widespread availability of noninvasive brain stimulation techniques such as transcranial magnetic stimulation (TMS) and transcranial direct and alternating current stimulation (TDCS, TACS), have led to a surge of interest in the use of electromagnetic neuromodulation to engage and manipulate brain oscillations, as a means of enhancing cognitive function and improving treatment of central nervous system disorders [21, 22]. Painless, inexpensive, and relatively easy to use, TMS, TDCS, and TACS protocols have already become popular to support a wide variety of clinical interventions [23–25]. These trends are most apparent in the fields of stroke rehabilitation [26, 27] and treatment of depression [28, 29], with promising results also being obtained in Alzheimer's [30, 31], Parkinson's [32], and epileptic [33, 34] cohorts. Noninvasive stimulation has also been used to interfere with brain oscillations in order to enhance working memory [35] and other cognitive functions (see [22] and references therein) and has been proposed as a tool to help slow down cognitive decline in the elderly [12].

Collectively, these results hold great promise. However, despite the rapid growth in recent years of applied and basic research in this area, the mechanisms by which noninvasive stimulation interferes with brain oscillations to modify neural and cognitive function remain poorly understood. This knowledge gap constitutes a major limitation in the development of new paradigms aimed at enhancing and/or restoring healthy brain dynamics through selective manipulation of synchrony and oscillatory activity. Mathematical and computational approaches represent increasingly central tools in efforts by scientists and clinicians to address these questions. For example, appropriately formulated mathematical models allow us to estimate and quantify the relationship between stimulation parameters (e.g. waveform, frequency, and intensity) and observed neural responses, at both the individual cell and neural population scales. These relationships can then be used to tune stimulation parameters in order to achieve the desired output in a clinical setting. In this chapter, we shall review some of the existing science about the effects of stimulation on brain rhythms and examine some of the mechanisms involved.

18.2 Brain Stimulation: Insights from Electrophysiology, Imaging, and Computational Models

From the perspective of individual cells, most information we have about how neurons (cortical and non-cortical) respond to stimulation comes from slice preparation studies [36]. The influence of static and varying electric fields on both extra- and intracellular potentials have been studied extensively in recent years, notably using computational and forward models [37, 38] and biophysical models [39, 40]. As a general rule, neurons subjected to varying electric fields will display fluctuations in membrane polarization mirroring the waveform applied. However, the connection between stimulation waveform and neural spiking activity is more challenging to establish and depends on many parameters that are difficult to determine in practice [41]. Weak electrical stimulation, such as that delivered to the brain by TDCS, cannot usually make neurons fire. Rather, it generates a diffuse electrical current that influences the firing probability by either hyperpolarizing or depolarizing neuronal membranes, hence causing subthreshold voltage fluctuations. Other forms of stimulation (e.g. direct cortical stimulation) do induce suprathreshold effects due to the greater proximity of the electrodes and more intense fields. Periodic stimulation triggers an alternation of depolarization and hyperpolarization periods across neuronal compartments [42], whose magnitude and phase depend on the relative position and orientation of the cell (and its arborization) with respect to the cathode/anode, as well as the frequency of the carrier signal. Due to synaptic delays and current leakage, neurons also exhibit filtering properties that make them more susceptible to low-frequency inputs if intensity is kept constant [41]. Some of these synaptic filtering effects are responsible for cell type-specific frequency tuning and resonances that have been suggested to direct the flow of information across the cortex [43, 44].

Macro-scale brain oscillations, such as the high-amplitude and low-frequency alpha rhythms that dominate human EEG and MEG recordings, are a collective phenomenon. It is therefore critical to understand how stimulation impacts populations and circuits, as opposed to individual neurons. Experimental studies have repeatedly demonstrated the ability of periodic brain stimulation to engage specific brain circuits in a frequency-specific manner, as well as its ability to both up- and downregulate performance on cognitive tasks [2, 45, 46].

The literature on computational models of brain stimulation is vast and spans the full range of spatial scales and neuroscience measurement modalities, from individual cells to whole-brain networks. For present purposes we focus on the meso-/macro-scale commensurate with both the brain stimulation techniques and recording techniques available in human subjects. Studies of microcircuits of recurrently connected excitatory and inhibitory spiking neurons [47, 48] and coupled bistable networks [49, 50] have demonstrated that periodic stimulation can evoke persistent large-scale oscillatory activity over prolonged time scales. Often however it is preferable for practical purposes to work with lower-dimensional neural population models (generally termed neural mass, mean field, or neural field models) than with detailed spiking networks, once the correspondence between simpler and more complex models has been established [51, 52]. Popular neural mass models such as those of Wilson and Cowan [53] or Jansen and Rit [54–56] represent entire brain regions by one or several excitatory and one or several inhibitory populations in a cortical column. Researchers deploying neural mass models to study brain stimulation have tended to focus on two things: (i) modulation of intrinsic brain rhythms by periodic stimulation [48, 57–62] and (ii) propagation of stimulation-evoked responses away from stimulated areas to downstream cortical and subcortical locations [63]. The power and precision of these large-scale modelling approaches have been improved considerably in recent years by developments in data acquisition and analysis tools for noninvasive estimation of individual subjects' white matter connectivity structure and geometry using diffusion-weighted MRI tractography [58, 63–66].

18.3   Restoring Brain Synchrony: A Control Problem

How can stimulation be used to compensate for disease-related loss in neural synchrony [9]? This control problem naturally depends on the pathology concerned, the impact it has on brain oscillations, and the cellular populations involved. Many physiological mechanisms may lead to the same change in neural synchrony. Irrespective of these details, our goal remains to restore collective brain activity by enhancing the amplitude of brain oscillations, speeding them up, or even suppressing them (as for example in the case of Parkinson's disease).

External signals (control signals) can engage and interact with these intrinsic oscillations via several specific, mathematically well-defined mechanisms. We list the main ones here:

1. *Resonance.* Stimuli with frequencies close to the system's natural frequency amplify the power of intrinsic oscillations. The proximity of the stimulation frequency to the intrinsic system frequency, and/or its harmonics, will dictate the amplitude of the responses.
2. *Entrainment.* The system is frequency- and phase-locked to the stimulus. The neuron or network of neurons is said to be *driven* by the stimulation, and phase alignment between the system's activity and the input must be observed. Entrainment of a nonlinear system is a function of stimulation frequency and amplitude, and regions in parameter space where entrainment occurs are called *Arnold Tongues*.
3. *Nonlinear acceleration.* High-frequency and/or random stimuli interact with the system's nonlinearities to provoke a *shift in intrinsic frequency*.
4. *Destabilization.* Application of an anti-phasic signal or a strong irregular (i.e. noisy) waveform transiently or permanently destabilizes (and thereby suppresses) intrinsic oscillations, effectively preventing the system from establishing stable synchronous states. This suppression can occur through a variety of ways (bifurcations).

For a comprehensive discussion of these different mechanisms, see [48]. We now turn to a didactic example of a simple model for EEG brain rhythms and their response to periodic stimulation, demonstrating aspects of mechanisms 1 and 2 outlined above.

18.4 Case Study: Tuning Thalamocortical Alpha Oscillations

Alpha oscillations, a cortical rhythm of roughly 8–12 Hz in humans, have been shown to play a key role in a wide variety of cognitive and perceptual functions [8, 67, 68] and, for that reason, have been a prime target for noninvasive stimulation [48]. To get some insights about the effect of patterned stimulation on alpha oscillations, let us consider the simplified thalamocortical network model shown diagrammatically in Fig. 18.1.

Numerous experimental and computational studies have shown that alpha oscillations are, least in part, of thalamocortical origin [69–71]. As such, in this nonlinear neural oscillator model, the dynamics are governed by a combination of cortico-cortical interactions and thalamocortical feedback. A similar network has recently been used to explain some state-dependent effects of repetitive stimulation on cortical activity in humans [62], and similar models have been used extensively by other authors [59, 72–76].

The basic building block of the model is a minimal four-component cortico-thalamocortical motif that describes the interconnections between excitatory ($\mathbf{u}_e$) and inhibitory ($\mathbf{u}_i$) neuronal populations in a given patch of cortical tissue and two thalamic nuclei: the reticular nucleus ($\mathbf{u}_r$) and specific relay nucleus ($\mathbf{u}_s$). Both relay and reticular nuclei receive inputs from the cortical excitatory population, following a corticothalamic conduction delay (τ_1). However only the relay nucleus sends excitatory input back to the cortex, again received following a corticothalamic delay. The reticular nucleus, which is widely known to have an inhibitory influence on other thalamic regions, plays a similar role to the cortical inhibitory population, inhibiting the relay nucleus and thereby generating oscillatory dynamics. Multiple adjacent patches of cortical tissue are modelled by repeating this basic circuit motif and coupling patches together through their excitatory neuronal populations. The

Fig. 18.1 *Thalamocortical model*. Schematic of the thalamocortical model structure. Cortical ($\mathbf{u}_e$,$\mathbf{u}_i$) and thalamic ($\mathbf{u}_s$,$\mathbf{u}_r$) populations interact through a delayed feedback loop. Entrainment of the network activity through electromagnetic stimulation P applied to $\mathbf{u}_e$ depends on the amplitude and frequency of the stimulation pulse, as well as the network state, controlled by I_o

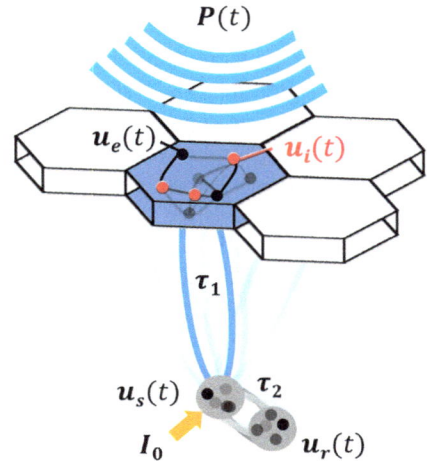

dynamics for the model are given by the following set of Wilson-Cowan-type stochastic delay-differential equations:

$$\frac{d\mathbf{u}_e(t)}{dt} = -\mathbf{u}_e(t) + g_{ee}\mathbf{F}[\mathbf{u}_e(t)] + g_{ei}\mathbf{F}[\mathbf{u}_i(t)] + g_{es}\mathbf{F}[\mathbf{u}_s(t - \tau_1)] + g_{cc}CCI$$

$$+ \mathbf{P}(t) + \sqrt{2D}\boldsymbol{\xi}_e(t) \tag{18.1}$$

$$\frac{d\mathbf{u}_i(t)}{dt} = -\mathbf{u}_i(t) + g_{ii}\mathbf{F}[\mathbf{u}_i(t)] + g_{ie}\mathbf{F}[\mathbf{u}_e(t)] + g_{is}\mathbf{F}[\mathbf{u}_s(t - \tau_1)] + \sqrt{2D}\boldsymbol{\xi}_i(t)$$

$$\tag{18.2}$$

$$\frac{d\mathbf{u}_r(t)}{dt} = -\mathbf{u}_r(t) + g_{re}\mathbf{F}[\mathbf{u}_e(t - \tau_1)] + g_{rs}\mathbf{F}[\mathbf{u}_s(t - \tau_2)] + \sqrt{2D}\boldsymbol{\xi}_r(t)$$

$$\tag{18.3}$$

$$\frac{d\mathbf{u}_s(t)}{dt} = -\mathbf{u}_s(t) + g_{sr}\mathbf{F}[\mathbf{u}_r(t - \tau_2)] + g_{se}\mathbf{F}[\mathbf{u}_e(t - \tau_1) + \sqrt{2D}\boldsymbol{\xi}_s(t) + I_o$$

$$\tag{18.4}$$

where the population vectors $\mathbf{u}_p(t) = (u_{p1}(t), u_{p2}(t), \ldots u_{pN}(t))$ denote the mean somatic membrane activity of neuronal population $p \in \{e, i, r, s\}$ and $N = 100$ units per population. Nonlinear interactions between neuronal populations are mediated by a steep sigmoidal response function $\mathbf{F}[u] = (1 + exp(-150))^{-1}$ and in the cases of intrathalamic and corticothalamic/thalamocortical connections are retarded by conduction delays $\tau_1 = 5$ ms and $\tau_2 = 20$ ms, respectively. The summed influence of the N excitatory populations in adjacent patches of cortical tissue on each other is given by the cortico-cortical input term $CCI = N^{-1}\sum_{k=1}^{N}\mathbf{F}[\mathbf{u}_e^k]$. Irregular and independent fluctuations are also present in the network, modelled by the zero mean Gaussian white noise processes ξ_p. Stimulation is applied evenly to all excitatory cortical units $\mathbf{u}_e$ through the continuous stimulation pulse $\mathbf{P} = M\sin(2\pi\omega t)$ with frequency ω and intensity M. Finally, the level of sensory drive to the thalamus is represented by the parameter I_o. As has been demonstrated previously [61], increasing this parameter past a critical point triggers suppression of resting state alpha oscillations and results in a greater susceptibility of cortical neural populations to entrainment by exogenous inputs or noninvasive stimulation.

Despite its simplicity, the base cortico-thalamocortical circuit motif in this model combines what are widely considered the two most important rhythmogenic mechanisms responsible for large-scale oscillatory activity observable in human extracranial recordings such as scalp EEG: low-frequency (i.e. alpha) oscillations generated by thalamocortical loops and higher-frequency (e.g. gamma) oscillations generated by local (intra-columnar) excitatory-inhibitory interactions in the cortex. Moreover, through variation of I_o, the model naturally accommodates the widely reported phenomenon of alpha suppression observed in response to sensory inputs [8, 77], during cognitive tasks requiring top-down control, and which is reduced in anaesthesia [78] and during sleep [73].

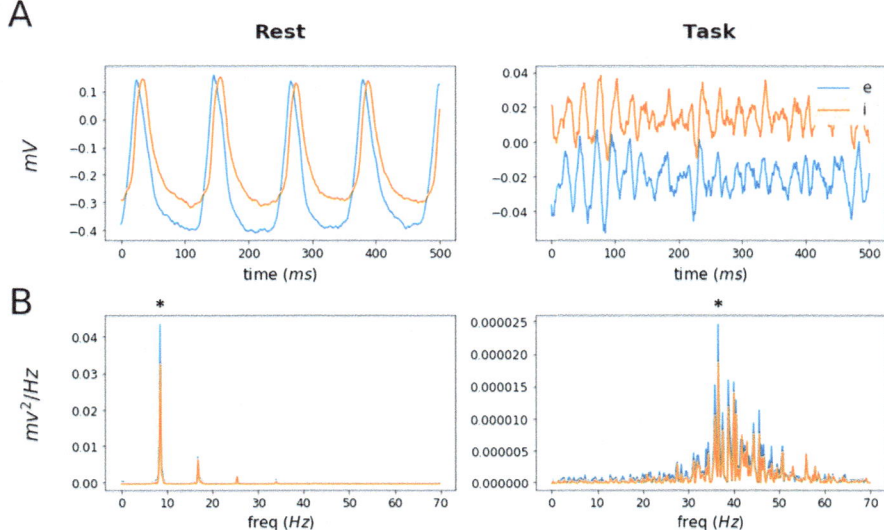

Fig. 18.2 *Baseline activity of the thalamocortical circuit in the rest and task states.* Time series (**a**) and power spectra (**b**) of cortical excitatory (blue) and inhibitory (orange) neural populations during rest (left column) and task (right column) states, in the absence of stimulation. The rest state ($I_o = 0$) is dominated by slow, high-amplitude alpha (around 10 Hz) rhythms, whereas the task state ($I_o = 1.5$) is dominated by lower amplitude gamma (around 40 Hz) oscillations. Asterisks indicate locations of spectral peaks

Figure 18.2 shows examples of the time series and power spectra generated by the model in 'rest' and 'task' states, in the absence of external stimulation ($\mathbf{P}(t) = 0$). The power spectrum in the rest state is dominated by high-amplitude, low-frequency (around 10 Hz) oscillations and their harmonics. As we shall see, this state-dependence of oscillatory activity has major implications for how we understand and model the effects of periodic brain stimulation. It has been shown that cortical systems are more prone to entrainment in regimes of strong sensory drive where the power of alpha is suppressed, suggestive of mechanisms such as stochastic resonance [61]. This has been repeatedly demonstrated in both noninvasive [79] and intracranial [62] experiments, highlighting the need to develop closed-loop systems where such fluctuations can be tracked – and compensated for – in real time. This dependence on baseline fluctuations is illustrated in Fig. 18.3, where the peak frequency and peak amplitude of the system in response to a sine wave stimulation are plotted as a function of stimulation amplitude and frequency. For both rest and task states, the lowest amplitude stimulation has no effect on the dominant rhythms, and so the plots are dominated by the endogenous natural frequency associated with that state (power spectral peaks from Fig. 18.2). As stimulation amplitude is increased, a triangular pattern known as an *Arnold Tongue* appears in panel A of Fig. 18.3. These patterns represent stimulation parameters for which

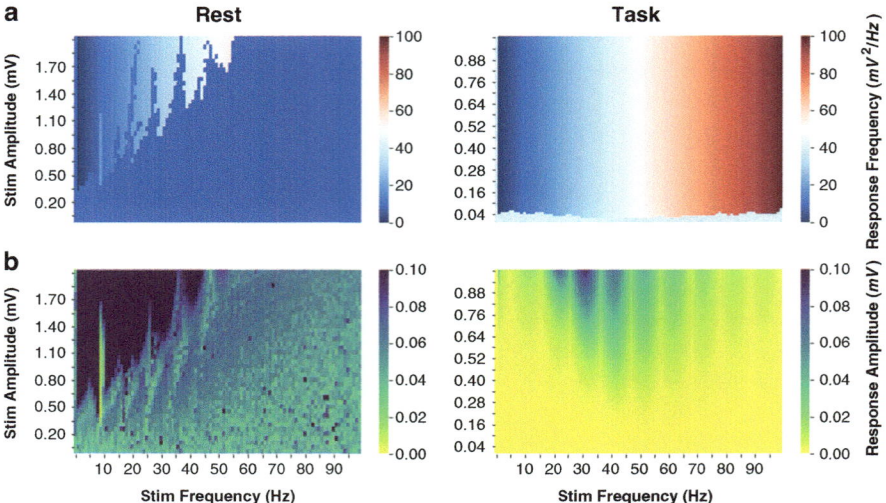

Fig. 18.3 *Entrainment region in both rest and task regimes.* Shown are maximum frequencies (**a**) and corresponding maximum amplitudes (**b**) of the cortical excitatory population in rest and task states as a function of stimulation frequency and stimulation amplitude. In both task and rest states, the system oscillates at its natural frequency (cf. Fig. 18.2, panel b), observable as a contiguous domain of constant dominant frequency in the lower portions of the frequency response heatmaps. Increasing stimulus level results in the emergence of Arnold Tongues (**a**) and resonance peaks (**b**), centred on the natural frequency

the system's activity is frequency- and phase-locked to the stimulus. Moreover, such susceptibility to entrainment is greatest at the system's natural frequency, which is around 10 Hz in the rest state and around 40 Hz in the task state (cf. Fig. 18.2, panel a). In addition to the difference in natural frequencies, the rest and task states differ markedly in their susceptibility to exogenous control. This can be seen by the fact that the stimulus level needed to entrain network oscillatory activity – i.e. the lower tip of the Arnold Tongues in Fig. 18.3 panel a – is much higher for the rest state than the task state. Resonance also shapes the amplitude of the responses, as seen in Fig. 18.3 panel b, where the amplitude of the solutions is increased near the system's endogenous frequency and/or its harmonics. Taken together, these results show that the presence of the alpha rhythm in the thalamocortical circuit can be understood as 'suppressing' higher-frequency exogenous oscillatory fluctuations, consistent with multiple empirical observations in the M/EEG literature.

18.5 Summary

In this chapter we have reviewed several key dynamical systems and control-theoretic concepts that, when combined with appropriate mathematical descriptions of macro-scale neural circuit properties, hold great promise in the development of precision brain stimulation therapies for neurological and neuropsychiatric disorders. The case study served to demonstrate several of these concepts in a schematic model exhibiting two distinct rhythmogenic mechanisms, both believed to play an important role in the genesis of macro-scale oscillations in the human brain: (i) intra-columnar excitatory-inhibitory interactions within the cortex and (ii) recurrent, delayed inhibitory feedback in thalamocortical loops. For the purposes of this chapter, we have focused on providing the reader with a clear exposition of some of this system's behavioural repertoire, rather than, e.g. matching precisely power spectrum traces from specific TMS/TACS-EEG datasets. Details of the full model implementation – including the use of optimization routines to accurately fit individual patients' whole-head power spectra, electromagnetic forward models (both brain-to-sensor and stimulus-to-brain), whole-brain white matter connectivity (anatomical macro-connectome), and integration with the virtual brain modelling and neuroinformatics platform [80, 81] – was beyond the scope of this chapter and shall be the subject of future work. It is nevertheless essential when constructing complex, multicomponent models to first have a detailed characterization of the dynamic behaviour of the individual building blocks – which in our case is the base thalamocortical motif shown in Fig. 18.1. The scientific value of such models lies in their ability to capture, in a mathematically compact way, some essential properties of large-scale neural dynamics and thereby provide insight into both normal activity and potential for control with brain stimulation techniques.

18.6 Software Note

Python code and additional documentation for the simulations presented in this chapter is freely available online at www.github.com/Lefebvrelab/ShapingBrainRhythms.

References

1. Fries P, Reynolds JH, Rorie AE, Desimone R (2001) Modulation of oscillatory neuronal synchronization by selective visual attention. Science 291:1560–1563
2. Hanslmayr S, Matuschek J, Fellner MC (2014) Entrainment of prefrontal beta oscillations induces an endogenous echo and impairs memory formation. Curr Biol 24:904–909
3. Duarte IC, Castelhano J, Sales F, Castelo-Branco M (2016) The anterior versus posterior hippocampal oscillations debate in human spatial navigation: evidence from an electrocortico-graphic case study. Brain Behav 6:e00507
4. Jacobs J, Korolev IO, Caplan JB, Ekstrom AD, Litt B, Baltuch G, Fried I, Schulze-Bonhage A, Madsen JR, Kahana MJ (2010) Right-lateralized brain oscillations in human spatial navigation. J Cogn Neurosci 22:824–836
5. Jadi MP, Sejnowski TJ (2014) Cortical oscillations arise from contextual interactions that regulate sparse coding. Proc Natl Acad Sci 111:6780–6785

6. Başar (2012) A review of alpha activity in integrative brain function: Fundamental physiology, sensory coding, cognition and pathology. Int J Psychophysiol 86:1–24
7. Tallon-Baudry (2003) Oscillatory synchrony and human visual cognition. J Physiol-Paris 97:355–363
8. Mierau A, Klimesch W, Lefebvre J (2017) State-dependent alpha peak frequency shifts: experimental evidence, potential mechanisms and functional implications. Neuroscience 360:146–154
9. Uhlhaas PJ, Singer W (2006) Neural synchrony in brain disorders: relevance for cognitive dysfunctions and pathophysiology. Neuron 52:155–168
10. Vardy AN, van Wegen EE, Kwakkel G, Berendse HW, Beek PJ, Daffertshofer A (2011) Slowing of M1 activity in Parkinson's disease during rest and movement – an MEG study. Clin Neurophysiol 122:789–795
11. Demirtaş M, Falcon C, Tucholka A, Gispert JD, Molinuevo JL, Deco G (2017) A whole-brain computational modeling approach to explain the alterations in resting-state functional connectivity during progression of Alzheimer's disease. Neuroimage: Clin 16:343–354
12. Assenza G, Capone F, di Biase L, Ferreri F, Florio L, Guerra A, Marano M, Paolucci M, Ranieri F, Salomone G, Tombini M (2017) Oscillatory activities in neurological disorders of elderly: biomarkers to target for neuromodulation. Front Aging Neurosci 9:189
13. Kielar A, Deschamps T, Chu RK, Jokel R, Khatamian YB, Chen JJ, Meltzer JA (2016) Identifying dysfunctional cortex: dissociable effects of stroke and aging on resting state dynamics in MEG and fMRI. Front Aging Neurosci 8:40
14. Kielar A, Deschamps T, Jokel R, Meltzer JA (2016) Functional reorganization of language networks for semantics and syntax in chronic stroke: evidence from MEG. Hum Brain Mapp 37:2869–2893
15. Chu RK, Braun AR, Meltzer JA (2015) MEG-based detection and localization of perilesional dysfunction in chronic stroke. Neuroimage: Clin 8:157–169
16. Oshino S, Kato A, Wakayama A, Taniguchi M, Hirata M, Yoshimine T (2007) Magnetoencephalographic analysis of cortical oscillatory activity in patients with brain tumors: synthetic aperture magnetometry (SAM) functional imaging of delta band activity. Neuroimage 34:957–964
17. Baayen JC, de Jongh A, Stam CJ, de Munck JC, Jonkman JJ, Trenité DG, Berendse HW, van Walsum AM, Heimans JJ, Puligheddu M, Castelijns JA (2003) Localization of slow wave activity in patients with tumor-associated epilepsy. Brain Topogr 16:85–93
18. van Wijk BC, Willemse RB, Vandertop WP, Daffertshofer A (2012) Slowing of M1 oscillations in brain tumor patients in resting state and during movement. Clin Neurophysiol 123:2212–2219
19. Sarnthein J, Jeanmonod D (2008) High thalamocortical theta coherence in patients with neurogenic pain. Neuroimage 39:1910–1917
20. Sarnthein J, Stern J, Aufenberg C, Rousson V, Jeanmonod D (2006) Increased EEG power and slowed dominant frequency in patients with neurogenic pain. Brain 129:55–66
21. Romei V, Bauer M, Brooks JL, Economides M, Penny W, Thut G, Driver J, Bestmann S (2016) Causal evidence that intrinsic beta-frequency is relevant for enhanced signal propagation in the motor system as shown through rhythmic TMS. Neuroimage 126:120–130
22. Thut G, Bergmann TO, Fröhlich F, Soekadar SR, Brittain JS, Valero-Cabré A, Sack AT, Miniussi C, Antal A, Siebner HR, Ziemann U, Herrmann C (2017) Guiding transcranial brain stimulation by EEG/MEG to interact with ongoing brain activity and associated functions: a position paper. Clin Neurophysiol 128:843–857
23. Dayan E, Censor N, Buch ER, Sandrini M, Cohen LG (2013) Noninvasive brain stimulation: from physiology to network dynamics and back. Nat Neurosci 16:3422
24. Paulus W (2011) Transcranial electrical stimulation (tES – tDCS; tRNS, tACS) methods. Neuropsychol Rehabil 21:602–617
25. Glannon W (2013) Neuromodulation, agency and autonomy. Brain Topogr 27:46–54
26. Hummel F, Celnik P, Giraux P, Floel A, Wu WH, Gerloff C, Cohen LG (2005) Effects of non-invasive cortical stimulation on skilled motor function in chronic stroke. Brain 128:490–499

27. Hummel FC, Cohen LG (2006) Non-invasive brain stimulation: a new strategy to improve neurorehabilitation after stroke? Lancet Neurol 5:708–712
28. Pascual-Leone A, Rubio B, Pallardó F, Catalá MD (1996) Rapid-rate transcranial magnetic stimulation of left dorsolateral prefrontal cortex in drug-resistant depression. The Lancet 348:233–237
29. Schlaepfer TE, Bewernick BH (2013) Neuromodulation for treatment resistant depression: state of the art and recommendations for clinical and scientific conduct. Brain Topogr 27:12–19
30. Freitas C, Mondragón-Llorca H, Pascual-Leone A (2011) Noninvasive brain stimulation in Alzheimer's disease: systematic review and perspectives for the future. Exp Gerontol 46:611–627
31. Fried I (2016) Brain stimulation in Alzheimer's disease. J Alzheimer's Dis 54(2):789–791
32. Dinkelbach L, Brambilla M, Manenti R, Brem AK (2017) Non-invasive brain stimulation in Parkinson's disease: exploiting crossroads of cognition and mood. Neurosci Biobehav Rev 75:407–418
33. Fridley J, Thomas JG, Navarro JC, Yoshor D (2012) Brain stimulation for the treatment of epilepsy. Neurosurg Focus 32(3):E13
34. Goodman JH (2004) Brain stimulation as a therapy for epilepsy. Adv Exp Med Biol 548:239–247
35. Violante IR, Li LM, Carmichael DW, Lorenz R, Leech R, Hampshire A, Rothwell JC, Sharp DJ (2017) Externally induced frontoparietal synchronization modulates network dynamics and enhances working memory performance. eLife 6:e2201
36. Reato D, Rahman A, Bikson M, Parra LC (2013) Effects of weak transcranial alternating current stimulation on brain activity – a review of known mechanisms from animal studies. Front Hum Neurosci 7:687
37. Manola L, Holsheimer J, Veltink P, Buitenweg JR (2007) Anodal vs cathodal stimulation of motor cortex: a modeling study. Clin Neurophysiol 118:464–474
38. Łęski S, Lindén H, Tetzlaff T, Pettersen KH, Einevoll GT (2013) Frequency dependence of signal power and spatial reach of the local field potential. PLoS Comput Biol 9:e1003137
39. Schiff ND (2010) Recovery of consciousness after brain injury: a mesocircuit hypothesis. Trends Neurosci 33:1–9
40. Schmidt SL, Iyengar AK, Foulser AA, Boyle MR, Fröhlich F (2014) Endogenous cortical oscillations constrain neuromodulation by weak electric fields. Brain Stimul 7:878–889
41. Lindén H, Pettersen KH, Einevoll GT (2010) Intrinsic dendritic filtering gives low-pass power spectra of local field potentials. J Comput Neurosci 29:423–444
42. Radman T, Su Y, An JH, Parra LC, Bikson M (2007) Spike timing amplifies the effect of electric fields on neurons: implications for endogenous field effects. J Neurosci 2:3030–3036
43. Bastos AM, Litvak V, Moran R, Bosman CA, Fries P, Friston KJ (2015) A DCM study of spectral asymmetries in feedforward and feedback connections between visual areas V1 and V4 in the monkey. Neuroimage 108:460–475
44. Bastos AM, Usrey WM, Adams RA, Mangun GR, Fries P, Friston KJ (2012) Canonical microcircuits for predictive coding. Neuron 76:695–711
45. Cecere R, Rees G, Romei V (2015) Individual differences in alpha frequency drive crossmodal illusory perception. Curr Biol 25:231–235
46. Cecere R, Romei V, Bertini C, Làdavas E (2014) Crossmodal enhancement of visual orientation discrimination by looming sounds requires functional activation of primary visual areas: a case study. Neuropsychologia 56:350–358
47. Ali MM, Sellers KK, Fröhlich F (2013) Transcranial alternating current stimulation modulates large-scale cortical network activity by network resonance. J Neurosci 33:11262–11275
48. Herrmann CS, Murray MM, Ionta S, Hutt A, Lefebvre J (2016) Shaping intrinsic neural oscillations with periodic stimulation. J Neurosci 36:5328–5337
49. Chaudhuri R, Knoblauch K, Gariel MA, Kennedy H, Wang XJ (2015) A large-scale circuit mechanism for hierarchical dynamical processing in the primate cortex. Neuron 88:419–431
50. Chaudhuri R, Bernacchia A, Wang XJ (2014) A diversity of localized timescales in network activity. eLife 3:e01239

51. Deco G, Ponce-Alvarez A, Mantini D, Romani GL, Hagmann P, Corbetta M (2013) Resting-state functional connectivity emerges from structurally and dynamically shaped slow linear fluctuations. J Neurosci 33:11239–11252
52. Stefanescu RA, Jirsa VK (2008) A low dimensional description of globally coupled heterogeneous neural networks of excitatory and inhibitory neurons. PLoS Comput Biol 4:31000219
53. Wilson HR, Cowan JD (1972) Excitatory and inhibitory interactions in localized populations of model neurons. Biophys J 12:1–24
54. Jansen BH, Rit VG (1995) Electroencephalogram and visual evoked potential generation in a mathematical model of coupled cortical columns. Biol Cybern 73:357–366
55. Valdes PA, Jiménez JC, Riera J, Biscay R, Ozaki T (1999) Nonlinear EEG analysis based on a neural mass model. Biol Cybern 81:415–424
56. Lopes da Silva FH, Hoeks A, Smits H, Zetterberg LH (1974) Model of brain rhythmic activity. Kybernetik 15:27–37
57. Spiegler A, Knösche TR, Schwab K, Haueisen J, Atay FM (2011) Modeling brain resonance phenomena using a neural mass model. PLoS Comput Biol 7:e1002298
58. Kunze T, Hunold A, Haueisen J, Jirsa V, Spiegler A (2016) Transcranial direct current stimulation changes resting state functional connectivity: a large-scale brain network modeling study. Neuroimage 140:174–187
59. Cona F, Lacanna M, Ursino M (2014) A thalamo-cortical neural mass model for the simulation of brain rhythms during sleep. J Comput Neurosci 37:125–148
60. Cona F, Zavaglia M, Massimini M, Rosanova M, Ursino M (2011) A neural mass model of interconnected regions simulates rhythm propagation observed via TMS-EEG. Neuroimage 57:1045–1058
61. Lefebvre J, Hutt A, Fröhlich F (2017) Stochastic resonance mediates the state-dependent effect of periodic stimulation on cortical alpha oscillations. eLife 6:e32054
62. Alagapan S, Schmidt SL, Lefebvre J, Hadar E, Shin HW, Fröhlich F (2016) Modulation of cortical oscillations by low-frequency direct cortical stimulation is state-dependent. PLOS Biol 14:e1002424
63. Spiegler A, Hansen EC, Bernard C, McIntosh AR, Jirsa VK (2016) Selective activation of resting-state networks following focal stimulation in a connectome-based network model of the human brain. eNeuro:3
64. Proix T, Spiegler A, Schirner M, Rothmeier S, Ritter P, Jirsa VK (2016) How do parcellation size and short-range connectivity affect dynamics in large-scale brain network models? Neuroimage 142:135–149
65. Deco G, Hagmann P, Hudetz AG, Tononi G (2013) Modeling resting-state functional networks when the cortex falls asleep: local and global changes. Cereb Cortex 24:3180–3194
66. Deco G, Senden M, Jirsa V (2012) How anatomy shapes dynamics: a semi-analytical study of the brain at rest by a simple spin model. Front Comput Neurosci 6:68
67. Klimesch W, Sauseng P, Hanslmayr S (2007) EEG alpha oscillations: the inhibition-timing hypothesis. Brain Res Rev 53:63–88
68. Klimesch W, Hanslmayr S, Sauseng P, Gruber WR, Doppelmayr M (2007) P1 and traveling alpha waves: evidence for evoked oscillations. J Neurophysiol 97:1311–1318
69. Hughes SW, Crunelli V (2005) Thalamic mechanisms of EEG alpha rhythms and their pathological implications. Neurosci 11:357–372
70. Hughes SW, Lörincz M, Cope DW, Blethyn KL, Kékesi KA, Parri HR, Juhász G, Crunelli V (2004) Synchronized oscillations at α and θ frequencies in the lateral geniculate nucleus. Neuron 42:253–268
71. Lörincz ML, Kékesi KA, Juhász G, Crunelli V, Hughes SW (2009) Temporal framing of thalamic relay-mode firing by phasic inhibition during the alpha rhythm. Neuron 63:683–696
72. Rennie CJ, Robinson PA, Wright J (2002) Unified neurophysical model of EEG spectra and evoked potentials. Biol Cybern 86:457–471
73. Robinson PA, Rennie CJ, Wright JJ, Bahramali H, Gordon E, Rowe DL (2001) Prediction of electroencephalographic spectra from neurophysiology. Phys Rev E 63:021903

74. Robinson PA, Rennie CJ, Wright JJ (1997) Propagation and stability of waves of electrical activity in the cerebral cortex. Phys Rev E 56:826–840
75. Robinson PA, Rennie CJ, Rowe DL, O'Connor SC, Gordon E (2005) Multiscale brain modelling. Philos Trans R Soc B: Biol Sci 360:1043–1050
76. Victor JD, Drover JD, Conte MM, Schiff ND (2011) Mean-field modeling of thalamocortical dynamics and a model-driven approach to EEG analysis. Proc Natl Acad Sci 108:15631–15638
77. Berger H (1933) Über das Elektrenkephalogramm des Menschen. Archiv für Psychiatrie und Nervenkrankheiten 98:231–254
78. Supp GG, Siegel M, Hipp JF, Engel AK (2011) Cortical hypersynchrony predicts breakdown of sensory processing during loss of consciousness. Curr Biol 21:1988–1993
79. Notbohm A, Kurths J, Herrmann CS (2016) Modification of brain oscillations via rhythmic light stimulation provides evidence for entrainment but not for superposition of event-related responses. Front Hum Neurosci 10:10
80. Ritter P, Schirner M, McIntosh AR, Jirsa VK (2013) The virtual brain integrates computational modeling and multimodal neuroimaging. Brain Connect 3:121–145
81. Sanz-Leon P, Knock SA, Spiegler A, Jirsa VK (2015) Mathematical framework for large-scale brain network modeling in the virtual brain. Neuroimage 111:385–430

Chapter 19
Brain Connectivity Reduction Reflects Disturbed Self-Organisation of the Brain: Neural Disorders and General Anaesthesia

Axel Hutt

Abstract The neurophysiological correlate of functional neural impairment is an open problem. Functional impairment may be observed as mental disorder, seizures or modification of consciousness level. The latter include loss of responsiveness under general anaesthesia, sleep or even trance in hypnosis. This chapter points out the relation between reduced brain connectivity as a possible correlate of neural functional impairment and self-organisation in the brain. A first numerical example demonstrates how neural noise disturbs self-organisation in the brain. Estimators of self-organisation such as global phase synchrony or information transfer quantify the degree of self-organisation. The chapter provides a brief literature review on how these estimators indicate brain connectivity modifications in neural disorders and under general anaesthesia.

Keywords Unconsciousness · Alzheimer's disease · Parkinson disease · Multiple sclerosis · Noise-induced transition

19.1 Introduction

The healthy normal brain can be regarded as an optimally tuned self-organised complex system [1–3]. It decodes sensory stimuli and encodes them to trigger responsive action. Multiple functional areas are known to transfer and share information. These properties result from a very high degree of self-organisation in and between functional areas, whose interactions enable the brain to process information. If these interactions are disturbed, then the brain can exhibit abnormal functions. Dependent on the degree of disturbed interactions, these abnormal functions are observed clinically as abnormal behaviour or even pathologies. The present chapter

A. Hutt (✉)
Department FE 12 (Data Assimilation), Deutscher Wetterdienst – German Meteorological Service, Research and Development, Offenbach am Main, Germany

© Springer Nature Switzerland AG 2019
V. Cutsuridis (ed.), *Multiscale Models of Brain Disorders*, Springer Series in Cognitive and Neural Systems 13, https://doi.org/10.1007/978-3-030-18830-6_19

discusses diseases and reduced consciousness emerging under drug administration, e.g. in hospital under general anaesthesia. The chapter reviews the role of brain connectivity in neural disorders and demonstrates how brain network fragmentation may explain loss of behavioural responsiveness in patients under general anaesthesia and why this reflects the partial breakdown of brain self-organisation.

19.2 Self-Organisation in the Brain

Complex systems exhibit a hierarchical structure of subunits that interact with each other [4, 5]. For weak interactions, the system dynamics is more or less given by the sum of the single subunits. However, for stronger interactions, subunits merge and generate new subunits on a higher hierarchical level. These higher-level units show dynamical behaviour that is not the sum of the subunits it emerged from but have new properties. Again, these new subunits on a higher hierarchical level interact with other new subunits generating together units on even higher hierarchical levels and so on. This merge of subunits is called self-organisation. The higher-level subunits can be observed as cooperative phenomena, such as cognitive functions [1], synchronisation [6, 7] or motor behaviour [3]. The enhanced interactions between subunits may indicate merged subunits and self-organisation. Hence a reasonable approach to reveal underlying neural mechanisms is the data analysis of neural activity that aims to quantify and identify interactions between subunits.

On a microscopic scale, single neurons or small neuron populations represent subunits, and a well-established data analysis approach is to extract synchronisation measures between these subunits from experimental data. For instance, enhanced synchronisation between single visual cortex neurons in visual perception tasks indicated cooperative interactions [8, 9]. This leads to the hypothesis [6] that the brain solves the visual binding problem by synchronisation or self-organisation.

On a macroscopic scale applying spatial mode analysis of electroencephalographic data [10–17] extracts spatial patterns that are supposed to reflect underlying interacting subunits. For instance, it has been shown that the spatio-temporal dynamics of middle-latent auditory evoked potentials is low-dimensional reflecting highly ordered neural activity [13]. This indicates neural self-organisation on a larger spatial and temporal scale.

To understand neural mechanisms and quantify the degree of self-organisation, brain connectivity has attracted much attention in recent years [18, 19]. For instance, [20] distinguished structural, effective and functional connectivity describing connections by fibre pathways, by correlations and by information flow, respectively. Other approaches aim to quantify influences of model variables [21] or brain regions or systems [22] on each other, e.g. by computing correlation coefficients or spectral or phase coherence.

19.2.1 An Illustrative Example

For illustration of self-organisation and how this can be quantified, let us consider a sparsely connected network of $N = 100$ nodes that is driven by additive random noise

$$dV_i = \left(-V_i + \frac{g}{N} \sum_{j=1}^{N} W_{ij} S[V_j(t - \tau)] \right) dt + \alpha \xi(t) \quad , \quad i = 1, \ldots, N \quad (19.1)$$

with $\alpha = 170/\sqrt{60}, g = 0.02$, the nonlinear transfer function S and Gaussian zero-mean i.i.d noise with $\langle \xi(t)\xi(t') \rangle = \mu^2 \delta(t - t')$. The parameter $\tau = 8$ ms is a delay time, and μ denotes the noise strength of the driving force. The network is sparse with connectivity probability 0.8 and the connection strength $W_{ij} = 1 \; \forall \; |i - j| \le 4$, $W_{ij} = -1$ otherwise, i.e. the network exhibits local excitation and lateral inhibition. Previous studies [23, 24] show that additive noise in such a network tunes the systems power spectrum and destructs coherent rhythmic activity for large noise levels. To gain the numerical solution of Eq. (19.1), we apply an Euler-Maruyama method [25] with discrete time step $\Delta t = 1$ ms.

Figure 19.1a shows the time-frequency distribution of the network mean $\bar{V}(t) = \sum_{i=1}^{N} V_i(t)/N$, and Fig. 19.1b gives the noise strength μ that changes with respect to time. We observe an oscillation of the network average with a single frequency for low noise levels and a destruction of this rhythm by additive noise. The different elements in the network are coherent at low noise levels as seen in Fig. 19.2, whereas coherence breaks down abruptly at larger noise levels. Since coherence reflects self-organisation in the network, we conclude that large noise destructs self-organisation. It is important to note that this destruction does not happen by weakening the direct coupling between elements but by rendering the interaction of network elements more noisy (see the discussion in [23, 24]).

Besides coherence measures, brain connectivity may be quantified by information theoretic measures [28, 29]. Figure 19.3 shows the transfer entropy (TE) [30] between two arbitrarily chosen time series $V_i(t)$ and $V_j(t)$ and the active information storage (AIS) [31] in $V_i(t)$ in different time windows. Since the noise level increases with time, Fig. 19.3 reveals that TE and AIS decrease with increasing noise level. Hence less information is passed between elements with increased noise level, and less information is stored in them. Since AIS reflects the degree of predictability from the corresponding time series, decreasing AIS is consistent with increased system randomness. Both measures may reflect the degree of self-organisation in the system, while typically TE is also interpreted as a brain connectivity measure.

The next section illustrates by selected literature examples that brain connectivity and neural disorders are strongly related. Then the subsequent section points out that modifications of brain connectivity during sedation and unresponsiveness under general anaesthesia resemble well findings in neural disorders. In sum,

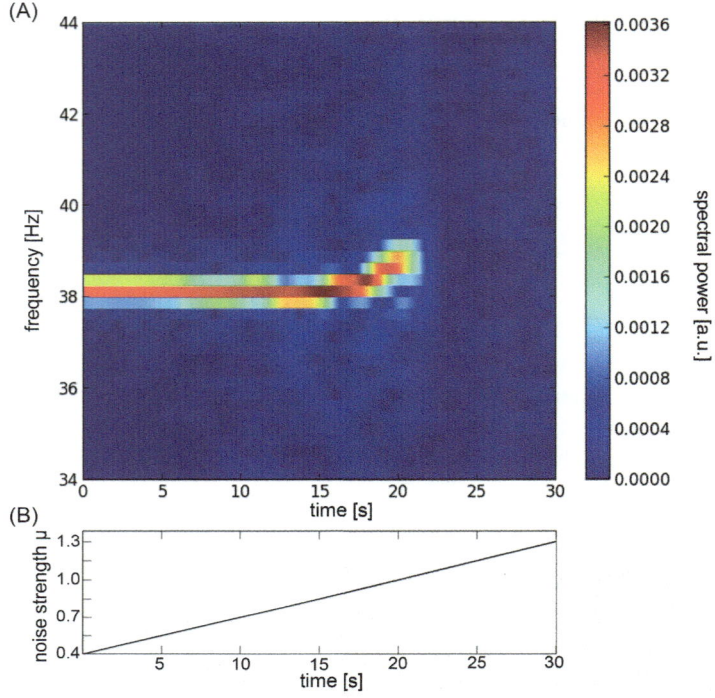

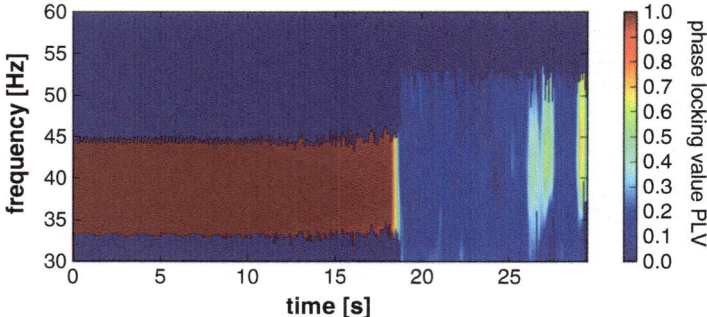

Fig. 19.1 Additive noise destructs coherent network activity. (**a**) Time-frequency distribution of the network mean $\bar{V}(t)$ gained from a Morlet wavelet analysis. (**b**) Noise strength μ with respect to time. The transfer function is chosen to $S(V) = 100/(1 + \exp(-100V))$

Fig. 19.2 Global phase locking for small noise and its destruction by large noise. This phase coherence is quantified by the phase-locking value PLV [26] and reflects the coherence in the system. Phase values are derived implicitly from wavelet transforms [26] and are defined for non-zero spectral only. To this end, we set PLV= 0 at time-frequency pairs where the spectral power $< 0.5 P_{max}$ with the maximum instantaneous spectral power P_{max} [27]

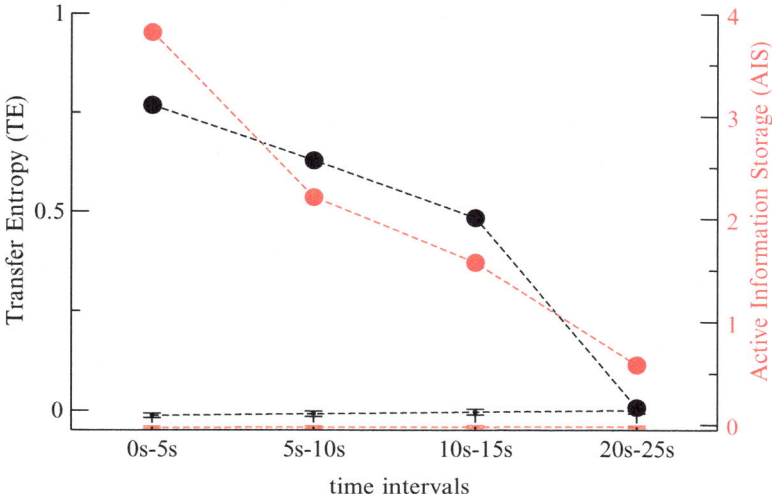

Fig. 19.3 Increasing noise decreases transfer entropy (TE) and active information storage (AIS) between and in two arbitrarily chosen network nodes, respectively. Thick dots mark the values for the signals; small dots with error bars represent results for 1000 surrogate data [28]. The estimates of TE and AIS have been computed with the open toolbox JIDT [28] with standard parameters and delay time $\tau = 8$ ms

this resemblance indicates that deterioration of connections (being structural or functional or effective) yields impairment of mental abilities as a consequence of a self-organisation deterioration.

19.3 Brain Connectivity in Neural Disorders

It is trivial to state that brain areas do not interact directly with each other if there are no fibre pathways between them. Consequently, in the following we assume present fibre pathways in the network if not stated otherwise. As already stated, enhanced brain connectivity may yield self-organisation. Such a well-balanced and self-organised brain state ensures normal information processing and good cognitive abilities. In turn weakened brain connectivity may yield abnormal functions, such as cognitive impairment, attentional deficits, unconsciousness or even diseases. This insight promises to contribute to early diagnostic examinations based on magnetic resonance imaging. Here the state of the art is the detection of structural abnormalities, e.g. in Alzheimer's disease [32], Parkinson disease [33] or multiple sclerosis [34].

Beyond the detection of structural pathologies, functional connectivity modifications appear to correlate well with structural pathologies and cognitive impairment.

For instance, multiple sclerosis (MS) patients are found to exhibit a functional fragmentation of the cerebellum [35] and the white matter integrity [36]. Interestingly, the latter study also shows enhanced functional connectivity in modules of the default mode network in early multiple sclerosis patients. The corresponding authors provide two explanations. The enhanced functional connectivity may represent a compensatory effect of the brain to solve given tasks that are more difficult for the brain with the loss of white matter integrity. Moreover, the finding could be explained by the loss of diversity in large-scale cortical dynamics due to white matter integrity loss. This missing ability of large-scale diverse cortical patterns may enhance local patterns' increasing local functional connectivity. This latter line of argumentation resembles the finding of a recent work on enhanced effective connectivity in local areas under general anaesthesia (see the subsequent section for more details).

These findings in multiple sclerosis resemble qualitatively in some respect to the brain connectivity in Alzheimer's disease (AD). Resting-state functional magnetic resonance imaging studies have revealed that AD patients exhibit a fragmentation of the brain network [37], e.g. within the default mode network [38] as in multiple sclerosis [36]. A further common feature to brain connectivity in multiple sclerosis patients is the enhanced functional connectivity in intralobe connections [37], such as in prefrontal lobe [39] or in parietal lobe [40].

In Parkinson disease (PD), dynamic functional brain deteriorations have been found in patients with mild cognitive impairment [41] and in idiopathic PD patients [42]. Similar to MS and AD, also enhanced correlations in certain brain areas have been found, e.g. in the prefrontal cortex [42] and in the default mode network [43].

Summarising, there is a strong indication that modification of functional connectivity is strongly related to brain disorders, be it reduction of functional connectivity in global networks or enhanced function connectivity in more local structures. Consequently, these disorders may result from strong reduction of network self-organisation.

19.4   Brain Connectivity Under General Anaesthesia

Arousal is an important action in the brain setting the level of excitation and hence controlling brain functions [44]. The ascending arousal system (ASS) [45] is a neural distributed network that sets the level of arousal via two major branches [46]. One branch is an ascending pathway from brainstem to the thalamus that inputs to the reticular nucleus and activates thalamic relay cells. Fast firing in cells in the upper brainstem areas, pedunculopontine and lateral tegmental nuclei, is present during wakefulness and rapid eye movement (REM) sleep, and low activity marks non-REM sleep stages. The other branch ascends, inter alia, from the brainstem and caudal hypothalamus and bypasses the thalamus projecting to the lateral hypothalamus, the basal forebrain and the cerebral cortex. Lesions along this pathway induce sleepiness and coma.

It is well-known that general anaesthetics alter neurotransmission in the cerebral cortex, thalamus and brainstem [47, 48]. Sedative and hypnotic anaesthetics diminish cortical activity in ferrets [49] and humans [50, 51]. This decrease of neural activity has been found by several experimental techniques, such as positron emission tomography (PET) as reduced metabolic activity [52], functional magnetic resonance imaging (fMRI) as reduced BOLD response and by electroencephalography (EEG) as reduced voltage amplitude [53]. For some general anaesthetics, this reduced activity can be explained by anaesthetics-enhanced inhibitory action of GABAergic receptors and anaesthetic-diminished excitatory action of NMDA receptors [54].

Taking a closer look at the AAS, its cortical and subcortical structures play different roles in arousal regulation. The brainstem projects along the two AAS pathways, activates or deactivates various AAS structures setting the excitation level of the cerebral cortex and controls respiratory and cardiovascular functions [48]. The brainstem is also supposed to largely influence thalamo-cortical oscillations that are observed in EEG [55]. A further major target structure in the AAS is the central thalamus that regulates the level of consciousness, and lesions in this area may produce neurological disorders of consciousness [56]. The hypothalamus may also promote awake state, while it primarily plays an important role in sleep regulation [46].

Anaesthetics affect the cortex and the functional structures in the AAS and primarily reduce their activity. This reduction may explain the interruption of information processing in the brain during anaesthesia. However, various experimental studies in the recent decade and the corresponding hypothesis of Tononi et al. [22, 57, 58] indicate that it is less the reduced neural activity that is the major marker of the anaesthetic state but rather the connectivity between neural structures. Various electrophysiological and modelling studies have shown that anaesthetics reduce the global connectivity in the brain [59] accompanied by a characteristic change of the activity power spectra [60–63]. These findings are partially consistent with experimental evidence from fMRI [64] revealing that global connectivity and the mean frequency of neural rhythms decrease with the level of sedation. A recent study on the information flow between ferret prefrontal cortex and visual cortex has revealed a decreasing transfer entropy between both areas under isoflurane sedation [65]. In the clinical context, patients with consciousness disorder show strong functional disconnections as well [64].

However this view is challenged by experimental evidence that, at surgery anaesthetic level that is deeper than for sedation, faster neural activity becomes more pronounced again [64], and hence the frequency shows a multiphasic relation to the level of anaesthesia. This multiphasic modification has been found in functional connectivity as well [66]. Some cortical and subcortical areas exhibit a first connectivity reduction, while increasing anaesthetic concentration before their connectivity was recovered at even deeper anaesthesia.

Moreover, anaesthetic sedation may even yield connectivity enhancement between certain structures, such as the increased connectivity between the precuneus and cortical areas in humans [67] or between the posterior cingulate cortex and, the sensorimotor cortices [68]. In an information entropy study of local field potentials in ferret prefrontal cortex and visual cortex, [65] found decreasing TE between prefrontal cortex and visual cortex, an increase in intra-area AIS and a decrease in intra-area entropy (H) with anaesthetic level in both areas. The authors of that study argue that less available information H in each area yields less information that can be transferred (TE). Synchronously, the increase of AIS reflects an increased predictability of the corresponding neural activity reflecting enhanced coherence. This is in line with the onset of coherence in the α−EEG frequency band at the point of loss of consciousness during propofol anaesthesia [69, 70].

A recent study of [71] further reveals that the onset of enhanced δ−power in EEG reflects well the point of loss of consciousness and that parietal and frontoparietal connectivity in this frequency range increases. Together with previous studies on EEG power spectra under general anaesthesia, this latter study indicates that connectivity may also depend on frequency range. Future studies will elicit whether brain connectivity is frequency-dependent.

In sum, this line of evidence suggests that an increased coherence in local areas by anaesthetic action induces the fragmentation of the global network. Huang et al. [64] bring up the hypothesis that, at first, sedation enhances local connections, while the global network remains unchanged, before global connections are reduced at deeper sedation levels. Then, in deep anaesthesia or in disorders of consciousness, local networks are fragmented as well.

19.5   Conclusion

Self-organisation in the brain is the condition for normal neural information processing in healthy patients. The reduction or removal of neural self-organisation is reflected, e.g. as neural disorders in diseases or loss of consciousness under anaesthesia. In these cases, the brain connectivity is affected in a similar manner. Local areas enhance their synchronisation and hence their connectivity, whereas global connections are reduced. These common features in neural disorders and anaesthesia (and even in sleep) point to the important role of connectivity in neural processing of information. Here, it is important to mention a line of evidence on the self-organisation in epileptic seizures [72]. It has been found experimentally that epileptic seizures exhibit high-frequency activity in local patches in the seizure onset zone [73], while desynchronisation has been observed at the seizure onset [74]. These, at a first glance contradictory findings, can be explained easily by spatial subsampling of electrodes detecting desynchronisation between low- and high-frequency spatial patches.

Future research will reveal whether these connections are strongly frequency-dependent. The distinction of connectivity in certain frequency bands may advance the understanding of diseases and general anaesthesia.

References

1. Singer W (1986) The brain as a self-organising system. Eur Arch Psychiatry Neurol Sci 236(1):4–9
2. Haken H (1996) Principles of brain functioning. Springer, Berlin
3. Kelso J (1995) Dynamic patterns: the self-organization of brain and behavior. MIT Press, Cambridge
4. Haken H (2004) Synergetics. Springer, Berlin
5. Nicolis G, Prigogine I (1977) Self-organization in non-equilibrium systems: from dissipative structures to order through fluctuations. Wiley, New York
6. Singer W (1993) Synchronisation of cortical activity and its putative role in information processing and learning. Annu Rev Physiol 55:349
7. Hutt A, Munk M (2006) Mutual phase synchronization in single trial data. Chaos Complex Lett 2(2):6
8. Koenig P, Engel A, Singer W (1995) Relation between oscillatory activity and long-range synchronization in cat visual cortex. Proc Natl Acad Sci USA 92:290–294
9. Singer W, Gray C (1995) Visual feature integration and the temporal correlation hypothesis. Ann Rev Neurosc 18:555–586
10. Uhl C, Kruggel F, Opitz B, von Cramon DY (1998) A new concept for EEG/MEG signal analysis: detection of interacting spatial modes. Hum Brain Map 6:137
11. Seifert B, Adamski D, Uhl C (2018) Analytical quantification of Shilnikov Chaos in epileptic EEG data. Front Appl Math Stat 4:57
12. Hutt A (2004) An analytical framework for modeling evoked and event-related potentials. Int J Bif Chaos 14(2):653–666
13. Hutt A, Riedel H (2003) Analysis and modeling of quasi-stationary multivariate time series and their application to middle latency auditory evoked potentials. Physica D 177:203
14. Fuchs A, Mayville J, Cheyne D, Einberg H, Deeke L, Kelso J (2000) Spatiotemporal analysis of neuromagnetic events underlying the emergence of coordinate instabilities. NeuroImage 12:71–84
15. Lehmann D, Skrandies W (1980) Reference-free identification of components of checkerboard-evoked multichannel potential fields. Electroenceph Clin Neurophysiol 48:609
16. Pascual-Marqui R, Michel C, Lehmann D (1995) Segmentation of brain electrical activity into microstates: model estimation and validation. IEEE Trans Biomed Eng 42(7):658–665
17. Hutt A, Schrauf M (2007) Detection of transient synchronization in multivariate brain signals, application to event-related potentials. Chaos Complex Lett 3(1):1–24
18. Jirsa V, McIntosh A (eds) (2007) Handbook of brain connectivity. Springer, New York
19. Sporns O (2010) Networks of the brain. MIT Press, Cambridge
20. Friston K (1994) Functional and effective connectivity in neuroimaging: a synthesis. Hum Brain Mapp 2:56–78
21. Friston K, Harrison L, Penny W (2003) Dynamic causal modeling. NeuroImage 19:466–470
22. Tononi G, Sporns O (2003) Measuring information integration. BMC Neurosci 4:31
23. Lefebvre J, Hutt A, Knebel J, Whittingstall K, Murray M (2015) Stimulus statistics shape oscillations in nonlinear recurrent neural networks. J Neurosci 35(7):2895–2903
24. Hutt A, Mierau A, Lefebvre J (2016) Dynamic control of synchronous activity in networks of spiking neurons. PLoS One 11(9):e0161488. https://doi.org/10.1371/journal.pone.0161488

25. Buckwar E, Kuske R, Mohammed S, Shardlow T (2008) Weak convergence of the Euler scheme for stochastic differential delay equations. LMS J Comput Math 11:60–99
26. Lachaux JP, Rodriguez E, Martinerie J, Varela F (1999) Measuring phase synchrony in brain signals. Human Brain Mapp 8:194–208
27. Hutt A, Lefebvre J, Hight D, Sleigh J (2018) Suppression of underlying neuronal fluctuations mediates EEG during general anaesthesia. Neuroimage 179:414–428 https://doi.org/10.1016/j.neuroimage.2018.06.043
28. Lizier J (2014) JIDT: an information-theoretic toolkit for studying the dynamics of complex systems. Front Robot AI 1:11
29. Wibral M, Vicente R, Lizier J (2014) Directed information measures in neuroscience. Springer, Berlin/Heidelberg
30. Wollstadt P, Martinez-Zarzuela M, Vicente R, Díaz-Pernas F, Wibral M (2014) Efficient transfer entropy analysis of non-stationary neural time series. PLoS One 9(7):e102833
31. Wibral M, Lizier J, Vögler S, Priesemann V, Galuske R (2014) Local active information storage as a tool to understand distributed neural information processing. Front Neuroinform 8:1
32. Scheltens P, Leys D, Barkhof F, Huglo D, Weinstein H, Vermersch P, Kuiper M, Steinling M, Wolters E, Valk J (1992) Atrophy of medial temporal lobes on MRI in "probable" Alzheimer's disease and normal ageing: diagnostic value and neuropsychological correlates. J Neurol Neurosurg Psychiatry 55:967–972
33. Yekhlef F, Ballan G, Macia F, Delmer O, Sourgen C, Tison F (2003) Routine MRI for the differential diagnosis of Parkinson's disease, MSA, PSP, and CBD. J Neural Transm 110(2):151–169
34. Sbardella E, Petsas N, Tona F, Pantano P (2015) Resting-state fMRI in MS: general concepts and brief overview of its application. Biomed Res Int 2015:212693
35. Dogonowski A, Andersen K, Madsen K, Sorensen P, Paulson O, Blinkenberg M, Siebner H (2014) Multiple sclerosis impairs regional functional connectivity in the cerebellum. Neuroimage: Clin 4:130–138
36. Hawellek D, Hipp J, Lewis C, Corbetta M, Engel A (2011) Increased functional connectivity indicates the severity of cognitive impairment in multiple sclerosis. Proc Natl Acad Sci USA 108(47):19066–19071
37. Wang K, Liang M, Wang L, Tian L, Zhang X, Li K, Jiang T (2007) Altered functional connectivity in early Alzheimer's disease: a resting-state fMRI study. Hum Brain Mapp 28:967–978
38. Greicius M, Srivastava G, Reiss A, Menon V (2004) Default-mode network activity distinguishes Alzheimer's disease from healthy aging: evidemce from functional MRI. Proc Natl Acad Sci USA 101:4637–4642
39. Horwitz B, Grady C, Schrageter N, Duara R, Rapoport S (1987) Intercorrelations of regional glucose metabolic rates in Alzheimer's disease. Brain Res 407:294–306
40. Bokde A, Lopez-Bayo P, Meindl T, Pechler S, Born C, Faltraco F, Teipel S, Moller H, Hampel H (2006) Functional connectivity of the fusiform gyrus during a face-matching task in subjects with mild cognitive impairment. Brain 129:1113–1124
41. Diez-Cirarda M, Strafella A, Kim J, Pena J, Ojeda N, Cabrera-Zubizarreta A, Ibarretxe-Bilbao N (2018) Dynamic functional connectivity in Parkinson's disease patients with mild cognitive impairment and normal cognition. NeuroImage: Clin 17:847–855
42. Hacker C, Perlmutter J, Criswell S, Ances B, Snyder A (2012) Resting state functional connectivity of the striatum in Parkinson's disease. Brain 135:3699–3711
43. Disbrowa E, Carmichael O, He J, Lanni K, Dressler E, Zhang L, Malhado-Chang N, Sigvardt K (2014) Resting state functional connectivity is associated with cognitive dysfunction in non-demented people with Parkinson's disease. J Parkinson's Dis 4:453–365
44. Quinkert A, Vimal V, Weil Z, Reeke G, Schiff N, Banavar J, Pfaff D (2011) Quantitative descriptions of generalized arousal, an elementary function of the vertebrate brain. Proc Natl Acad Sci USA 108:15617–15623
45. Moruzzi G, Magoun H (1949) Brainstem reticular formation and activation of the EEG. Electroencephalogr Clin Neurophysiol 1:455–473

46. Saper C, Scammell T, Lu J (2005) Hypothalamic regulatrion of sleep and circadian rhythms. Nature 437:1257–1263
47. Alkire M, Hudetz A, GTononi (2008) Consciousness and anesthesia. Science 322:876–880. https://doi.org/10.1126/science.1149213
48. Brown E, Lydic R, Schiff N (2010) General anesthesia, sleep, and coma. N Engl J Med 363:2638–2650
49. Sellers KK, Bennett DV, Hutt A, Frohlich F (2013) Anesthesia differentially modulates spontaneous network dynamics by cortical area and layer. J Neurophysiol 110:2739–2751
50. Purdon PL, Pierce ET, Mukamel EA, Prerau MJ, Walsh JL, Wong KF, Salazar-Gomez AF, Harrell PG, Sampson AL, Cimenser A, Ching S, Kopell NJ, Tavares-Stoeckel C, Habeeb K, Merhar R, Brown EN (2012) Electroencephalogram signatures of loss and recovery of consciousness from propofol. Proc Natl Acad Sci USA 110:E1142–1150
51. Murphy M, Bruno MA, Riedner BA, Boveroux P, Noirhomme Q, Landsness EC, Brichant JF, Phillips C, Massimini M, Laureys S, Tononi G, Boly M (2011) Propofol anesthesia and sleep: a high-density EEG study. Sleep 34(3):283–291
52. Fiset P, Paus T, Daloze T, Plourde G, Meuret P, Bonhomme V, Hajj-Ali N, Backman S, Evans A (1999) Brain mechanisms of propofol-induced loss of consciousness in humans: a positron emission tomographic study. J Neurosci 19(13):5506–5513
53. Purdon PL, Pierce E, Bonmassar G, Walsh J, Harrell G, Deschler D, Kwo J, Barlow M, Merhar R, Lamus C, Mullaly C, Sullivan M, Maginnis S, Skoniecki D, Higgins H, Brown EN (2009) Simultaneous electroencephalography and functional magnetic resonance imaging of general anesthesia. Ann NY Acad Sci 1157:61–70
54. Franks N, Lieb W (1994) Molecular and cellular mechanisms of general anesthesia. Nature 367:607–614
55. Scheib C (2017) Brainstem influence on thalamocortical oscillations during anesthesia emergence. Front Syst Neurosci 11:66
56. Schiff N (2008) Central thalamic contributions to arousal regulation and neurological disorders of consciousness. Ann NY Acad Sci 1129:105–118
57. Tononi G (2004) An information integration theory of consciousness. BMC Neurosci 5:42
58. Pillay S, Vizuete J, Liu X, Juhasz G, Hudetz A (2014) Brainstem stimulation augments information integration in the cerebral cortex of desflurane-anesthetized rats. Front Integr Neurosci 8:8
59. Hudetz A, Mashour G (2016) Disconnecting consciousness: is there a common anesthetic end point? Anesth Analg 123(5):1228–1240
60. Boly M, Moran R, Murphy M, Boveroux P, Bruno MA, Noirhomme Q, Ledoux D, Bonhomme V, Brichant JF, Tononi G, Laureys S, Friston KI (2012) Connectivity changes underlying spectral EEG changes during propofol-induced loss of consciousness. J Neurosci 32(20):7082–7090
61. Vizuete J, Pillay S, Ropella K, Hudetz A (2014) Graded defragmentation of cortical neuronal firing during recovery of consciousness in rats. Neuroscience 275:340–351
62. Lewis L, Weiner V, Mukamel E, Donoghue J, Eskandar E, Madsen J, Anderson W, Hochberg L, Cash S, Brown E, Purdon P (2012) Rapid fragmentation of neuronal networks at the onset of propofol-induced unconsciousness. Proc Natl Acad Sci USA 109(21):E3377–3386
63. Hashemi M, Hutt A, Sleigh J (2015) How the cortico-thalamic feedback affects the EEG power spectrum over frontal and occipital regions during propofol-induced anaesthetic sedation. J Comput Neurosci 39(1):155
64. Huang Z, Liu X, Mashour G, Hudetz A (2018) Timescales of intrinsic bold signal dynamics and functional connectivity in pharmacologic and neuropathologic states of unconsciousness. J Neurosci 38(9):2304–2317
65. Wollstadt P, Sellers K, Rudelt L, Priesemann V, Hutt A, Frohlich F, Wibral M (2017) Breakdown of local information processing may underlie isoflurane anesthesia effects. PLoS Comput Biol 13(6):e1005511

66. Liu X, Pillay S, Li R, Vizuete J, Pechman K, Schmainda K, Hudetz A (2013) Multiphasic modification of intrinsic functional connectivity of the rat brain during increasing levels of propofol. NeuroImage 83:581–592
67. Liu X, Li S, Hudetz A (2014) Increased precuneus connectivity during propofol sedation. Neurosci Lett 561:18–23
68. Stammatakis E, Aadapa R, Absalom A, Menon D (2010) Changes in resting neural connectivity during propofol sedation. PLoS One 5:e14224
69. Cimenser A, Purdon PL, Pierce ET, Walsh JL, Salazar-Gomez AF, Harrell PG, Tavares-Stoeckel C, Habeeb K, Brown EN (2011) Tracking brain states under general anesthesia by using global coherence analysis. Proc Natl Acad Sci USA 108(21):8832–8837
70. Supp G, Siegel M, Hipp J, Engel A (2011) Cortical hypersynchrony predicts breakdown of sensory processing during loss of consciousness. Curr Biol 21:1988–1993
71. Lee M, Sanders R, Yeom SK, Won D-O, Seo K, Kim H, Tononi G, Lee S (2017) Network properties in transitions of consciousness during propofol-induced sedation. Sci Rep 7:16791
72. Jiruska P, de Curtis M, Jefferys JGR, Schevon CA, Schiff SJ, Schindler K (2013) Synchronization and desynchronization in epilepsy: controversies and hypothesis. J Physiol 591(4):787–797. https://doi.org/10.1113/jphysiol.2012.239590
73. Schevon CA, Goodman RR, McKhann G Jr, Emerson RG (2010) Propagation of epileptiform activity on a submillimeter scale. J Clin Neurophysiol 27:406-411
74. Netoff TI, Schiff SJ (2002) Decreased neuronal synchronization during experimental seizures. J Neurosci 22:7297–7307

Index

A

Abbott, L.F., 22
Absence seizures, 168, 171–177, 179, 181, 182
Accumulator model with lateral inhibition, 91
Acetylcholine, 113–123
Action sequences, 67–76
Adams, R.A., 81–88
Ahn, S., 57–63
Akinesia, 43, 44, 58, 105
Alzheimer's, 113–123, 149–153, 194, 211, 212
Alzheimer's disease (AD), 113–123, 149–153, 211, 212
Antisaccade paradigm, 91, 100
Apathy, 127–133
ASD, *see* Autism spectrum disorder (ASD)
Astrocytes, 109–111
Attention, 22, 105, 109, 110, 114, 115, 117, 138, 140, 141, 143–145, 178, 208
Autism spectrum disorder (ASD), 135–145
Autobiographical memory (AM), 157–161
Avoli, M., 177

B

Balc, F., 67–76
Baldassarre, G., 127–133
Basal ganglia (BG), 4–6, 8, 11, 13–19, 30, 31, 33, 41–50, 58, 61–63, 106–108, 110, 136
Basal ganglia network model, 5, 14, 15, 19, 41–50
Bayesian, 82–84, 87
Beads task, 83–87
Berényi, A., 177

Beta-band oscillations, 49, 57–58, 63
Beta-range oscillation, 14, 19
Beta-rebound, 21–26
BG, *see* Basal ganglia (BG)
Bistability, 168, 171, 177
Bowman, H., 113–123
Bradykinesia, 4, 11, 29–37, 43–49, 58, 105
Brain fingerprints, 145
Breakspear M, 22
Brunel, N., 117
Bryson, S.E., 140
Byrne, Á., 21–26

C

CA1, 152, 153, 187–190
Caligiore, D., 127–133
Carpenter, R.H.S., 98, 100
Chatzikalymniou, A., 149–153
Cognitive modeling, 91–102
Computational, 101, 102, 122, 127–133, 136, 139–145, 159, 194, 195, 197
Computational modeling, 5, 26, 42, 43, 45, 46, 48, 49, 83, 106–111, 113–123, 127–129, 133, 144, 150–152, 168, 169, 171, 182, 186–188, 190, 194–196
Computational neuroscience, 24
Computer model, 167–182
Conway, M.A., 158
Coombes, S., 21–26
Counter-stimulation, 177
Courchesne, E., 144
Cowan, J.D., 171, 195
Cutsuridis, V., 29–37, 91–102, 151

© Springer Nature Switzerland AG 2019
V. Cutsuridis (ed.), *Multiscale Models of Brain Disorders*, Springer Series in
Cognitive and Neural Systems 13, https://doi.org/10.1007/978-3-030-18830-6

D
Deco, G., 115, 117, 118
Deep brain stimulation (DBS), 3–11, 41–50, 63
Depression, 26, 101, 108, 127, 128, 132, 139, 194
Destexhe, A., 170
Desynchronization, 47–49, 59, 60, 62, 63, 144
Dopamine (DA), 7, 9, 31–33, 35, 36, 42–44, 58, 63, 105–110, 127, 129, 130
Doudet, D.J., 34
Doya, K., 13–19
Duch, W., 135–145
Dynamical neural systems, 171–172

E
Entropy, 68, 70–75, 144, 214
Epilepsy, 115, 136, 167–182, 185–187, 189
Everling, S., 100
Excitation-inhibition balance, 22, 186
Eye movements, 82, 93, 94

F
Fast ripples, 187
Feynman, R., 182
Finkel, L.H., 151
Fiore, V.G., 127
Frei, M.G., 177

G
Gamma oscillations, 115, 189, 198, 199
Generalized epilepsies, 177
Gerstein, G.L., 174
Glass, L., 168
Globus pallidus, 4, 5, 14, 15, 30, 33, 42, 62, 107
Goodfellow, M., 181
Griffiths, J.D., 193–201
Grill, W.M., 41–50

H
Hélie, S., 105–111
High-frequency oscillations (HFOs), 186, 187
Hippocampus, 151–153, 161, 185–190
Huang, Z., 214
Huguenard, J.R., 178
Humphreys, G.W., 117
Hutt, A., 207–215

I
Igarashi, J., 13–19
I_h current, 171, 175, 176, 178, 182
Impulse control, 91
Information theory, 67–76
Inter-ictal and ictal distributions, 173, 174, 176, 182
Intermittency, 61, 62, 181
Izhikevich spiking neuron, 5, 6

J
Jansen, B.H., 195
Jardri, R., 87
Johnston, K., 100

K
Kalitzin, S., 167–182
Kawakubo, Y., 140
Klein, J.P., 180
Koppert, M.M.J., 175, 182
Kozák, G., 177
Kumaravelu, K., 41–50

L
Landry, R., 140
Lefebvre, J.R., 193–201
Lempel-Ziv complexity (LZC), 68–70, 72–75
Lesne, A., 68, 72
Leys, C., 69
Liddle, P.F., 21–26
Lopes da Silva, F.H., 167–182
Lüttjohann, A., 177
Lytton, W.W., 169

M
Mackey, M.C., 168
Mandali, A., 3–11
Mandelbrot, B., 174
Mataix-Cols, D., 76
Mavritsaki, E., 113–123
Mazaheri, A., 116, 123
Mean-field model, 24, 25
Meeren, H.K., 179
Memory loss, 157–161
Menschik, E.D., 151
Mental disorders, 68, 135
Miao, C., 157–161
Milton, J., 181

Model-based optimization, 42, 49
Monchi, O., 107
Morén, J., 13–19
Moroney, R., 11
Motor cortex, 14, 22, 23, 31, 32, 44
Moustafa, A.A., 3–11, 67–76, 157–161
Movement disorders, 42
Multiple sclerosis (MS), 212
Munakata, Y., 140

N
Nambu, A., 44
Na-persistent current, 170, 178, 182
Neural mass, 195
Neural mass models, 21–26, 151, 175, 177,
        180, 181, 195
Neural oscillations, 58, 115
Neural synchronization, 57–63
Neurocomputational model, 159, 160
Neurodynamics, 58, 135–145
Noise-induced transition, 181
Noorani, I., 100
Noradrenaline, 127

O
Obsessive-compulsive disorder (OCD), 68, 71,
        73, 76, 91–102, 115
O'Reilly, R.C., 140
Osorio, I., 177
Öztel, T., 67–76

P
Park, C., 57–63
Parkinson disease (PD), 3–11, 14–19, 29–37,
        41–50, 57–63, 105–111, 196, 211, 212
Paz, J.T., 178
Pierce, K., 144
Pleydell-Pearce, C.W., 158
Poil, S.-S., 115
Polack, P.O., 178
Psychosis, 81, 83

R
Raikov, I., 185–190
Rapoport, J.L., 76
Research Domain Criteria (RDoC), 136, 138
Response inhibition, 92

Rit, V.G., 195
Rubchinsky, L.L., 57–63
Rubio, S.E., 151

S
Sajedinia, Z., 105–111
Schizophrenia, 21–26, 81–88, 97, 100, 101,
        107
Schomburg, E., 189
Sejnowski, T.J., 170
Serotonin, 127–133
Shannon metric entropy, 68, 71–73
Shouno, O., 13–19
Silvetti, M., 127–133
Skinner, F.K., 149–153
Slowness of movement, 31, 35, 58
Soltesz, I., 185–190
Sorokin, J.M., 177, 178
Spiking neural network, 14–16, 19, 22
Spiking Search over Time and Space model,
        117
Spinal cord, 14, 30, 31
Srinivasa Chakravarthy, V., 3–11
STN, *see* Sub thalamic nucleus (STN)
Stuke, H., 87
Su, L., 113–123
Subcortical lesion, 42, 45
Sub-harmonic resonance, 14
Sub thalamic nucleus (STN), 4–9, 11, 14–16,
        42–50, 61, 62, 106, 107
Suffczynski, P., 167–182
Superior colliculus, 92

T
Tan, A.-H., 157–161
Thalamus, 4–9, 13–19, 30, 42, 48, 107, 169,
        172, 177, 179, 198, 212, 213
Theta oscillations, 116, 187
Thiele, A., 115, 117, 118
Tononi, G., 213
Tremblay, L., 36
Tremor, 3–11, 13–19, 43, 45, 48, 49, 58, 61,
        63, 105
Triphasic pattern of muscle activation, 31

U
Uddin, L.Q., 144
Unconsciousness, 211

V
Van Heukelum, S., 177
van Luijtelaar, G., 177
Verret, L., 151
Vogels, T.P., 22

W
Walton, M.E., 132
Wang, D., 157–161
Wang, X.J., 117, 169
Williams, M.L.L., 98

Wilson, H.R., 171, 195
Wuensch, K.L., 99

Y
Yener, G.G., 116
Yoshimoto, J., 13–19

Z
Zeki, M., 67–76
Zihl, J., 118
Zimmerman, A.W., 139

Printed by Printforce, the Netherlands